The Energy Construct
2nd Edition

Copyright © 2007 Ben Cipiti
All rights reserved.
ISBN: 1-4196-6978-8
Library of Congress Control Number: 2007903962

To order additional copies, please contact us.
Booksurge, LLC
www.booksurge.com
1-866-308-6235
orders@booksurge.com

Cover photo courtesy of NASA

The Energy Construct
2nd Edition

Achieving a Clean, Domestic,
and Economical Energy Future

Ben Cipiti

Contents

Preface vii

Introduction 1
    Chapter 1: The Energy Challenge 3

Part I: The Transportation Sector 17
    Chapter 2: Biofuels 19
    Chapter 3: Hydrogen 26
    Chapter 4: Electric Vehicles 33

Part II: Renewable Energy 43
    Chapter 5: Solar Energy 45
    Chapter 6: Geothermal Energy 58
    Chapter 7: Hydroelectric & Ocean Energy 68
    Chapter 8: Biomass Energy 77
    Chapter 9: Wind Energy 84

Part III: Nuclear Energy 95
    Chapter 10: Nuclear Energy 97
    Chapter 11: Fusion Energy 111

Part IV: Fossil Energy 125
    Chapter 12: Coal Energy 127
    Chapter 13: Natural Gas Energy 141

Conclusion 151
    Chapter 14: The Energy Construct 153

    Appendix 167
    References 169

Preface

FOR THE PAST THIRTY YEARS our country has become increasingly more aware of the effects that our energy use has on the world. The ideas of energy efficiency, conservation, and renewable energy have been around for a while now, yet why is it that we still get most of our energy from oil, coal, and natural gas? We know that we have a problem, but the challenges of moving away from the established technologies and infrastructure are monumental.

My goal in writing this book was to take a realistic look at how our country can achieve a better energy future. This problem is not just about developing clean energy sources—it is also about developing economical sources of energy so that they can compete in today's tight energy market. And it is about reducing our dependence on foreign sources of energy and finding solutions that will be supported by our local communities.

I know there are better options available. There are a number of fascinating clean energy options available today, but many factors prevent their widespread use. As an engineer, I tend to look at a solution to a problem and ask not if it is possible, but if it is feasible—in many ways this drives the perspective of this book. Whenever I learn about some new energy concept, I immediately want the tough questions answered. How much will it cost? How reliable will it be? Can it be used all over the country or will its use be limited? My experience has taught me to look at the whole picture.

There is a solution to our energy problem, and I hope this book provides a path forward. But this is a problem that our country will need to solve together. It is not just about technology; the problem also requires public support and political will. We need to be willing to make large-scale changes, and some may not be easy. Our energy supply drives so many aspects of our lives, and the decisions we make or do not make in the next few years could lead to drastic consequences to our economy, our security, and our world.

Introduction

ENERGY IS THE DRIVING FORCE of our world. In our modern lives, we are absolutely dependent on energy for everything from heating and powering our homes to satisfying our transportation needs. Our economy revolves around cheap and plentiful sources of fuel and electricity. This age of incredible advances in medicine, technology, and information resources is only possible with our constant supply of energy. But our population continues to expand, and energy demand keeps rising.

In the first chapter I discuss the challenge of meeting our energy needs. There is not an easy solution since a number of factors play a role in deciding how we will satisfy energy demand. However, understanding and appreciating all the factors is the only way that our country will be able to move forward.

Chapter 1: The Energy Challenge

THE GREATEST ENGINEERING CHALLENGE of this century will be to satisfy our voracious appetite for energy resources with clean, safe, domestic, and affordable technologies. The climate change debate continues to rage on with strong agendas on both sides of the issue. Yet despite the debate, we are seeing how clean technologies are beginning to win out. Increasingly the dirty energy technologies are paying more in the long term for clean up and mitigation. Those technologies that can minimize environmental impact see more acceptability and less long-term risk. But which of the clean energy technologies can really compete?

Our dependence on foreign sources of oil to power the transportation sector continues to be an area of concern. The conflicts in the Middle East raise many questions about a war's worth. The desire to spread democracy and the desire to prevent terrorism are worthy goals, but how much exactly does the oil factor play a role? How much of the billions of dollars spent on the wars of the past several years is justified by securing energy resources? Our country needs domestic sources of energy for the transportation sector to prevent future threats to our security.

Plenty of clean and domestic alternative energy technologies do exist, but we also need economical options. A small subset of the population may be willing to pay more for clean energy, but it is important to keep in mind that many Americans have a hard time paying their current utility and gas bills. Clean energy alternatives will never see widespread use unless the costs are competitive. Government regulations or incentives can help to make clean energy options more economical, but the development path must eventually make the technology competitive without subsidies.

Where do we stand now? The top pie chart in Figure 1 shows the percent breakdown of our total primary energy use—this includes all energy used for electricity generation, transportation fuels, heating, and industrial processes.[1] It is obvious from this figure how heavily dependent we are on fossil fuels. Our continued dependence on oil, coal, and natural gas releases a tremendous amount of greenhouse gases. To reduce emissions, our country must focus on these dominant sources of energy.

The bottom pie chart in Figure 1 shows the percent breakdown of only the electric generation sector. Electricity generation accounts for about 41% of our total energy use, so the bottom chart is just a subset of the top chart. Coal, natural gas, and nuclear dominate electrical production. The renewable energy options make a small, but growing contribution. These two graphs may be useful to refer back to while reading the book.

Our population is increasing, which means our energy demand is increasing. Current projections suggest a 0.5% increase in total energy consumption per year and a 1.0% increase in electrical demand per year.[2] To satisfy this increase in demand, our country will need to build approximately 120 large power plants in the next 20 years. How are we going to satisfy increasing demand while at the same time reduce emissions?

The purpose of this book is to take a realistic look at achieving a clean, domestic, and economical energy future. This book is not about small-scale changes. Instead it is about tackling the dominant problem areas of energy use: burning fossil fuels for transportation and power generation.

The Energy Challenge

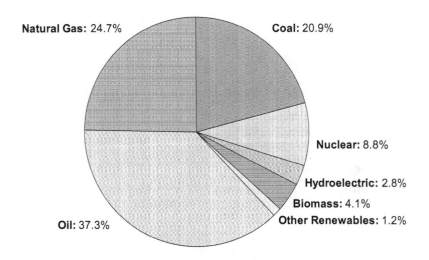

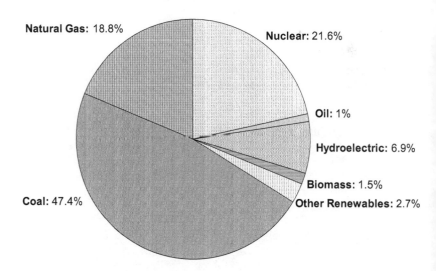

Figure 1: U.S. Energy Use[1]

I examine a number of power generation and transportation technologies throughout this book from different perspectives including environmental impact, economics, the domestic resource base, public acceptability, and reliability. The goal is to find those technologies that are likely to make dramatic, near-term differences in the way we generate and use energy.

Units of Energy and Power

Since a great deal of technology is discussed throughout this book, it is important to understand the units of energy and power that are used. A Joule (J) is a measure of energy that is used often throughout the book. In the case of fossil fuels, energy content refers to the amount of heat produced by burning the fuel. A Watt (W), on the other hand, is a unit of power and represents the amount of energy produced per unit of time. One watt is equal to the production of one joule per second:

$$1 \text{ W} = 1 \text{ J/s}$$

In any discussion on energy or power production on the large scale, these units are scaled up using the following prefixes:

kilowatt (kW)	1,000 watts
megawatt (MW)	1,000,000 watts
gigawatt (GW)	1,000,000,000 watts

Likewise, the same prefixes can be used for scaling up joules. The focus of this book is electrical generation from power plants, typically rated in terms of megawatts (MW).

A further complication is that there is a difference between the amount of thermal power a plant generates and the amount of useful electric power generated. The conversion efficiency is the efficiency for transforming a thermal energy source into electrical power. For most steam-driven power plants the conversion efficiency may vary between 35-45%. The rest of the energy is dumped into the atmosphere or environment as waste heat. This may appear to be a low efficiency, but laws of thermodynamics limit how much energy can be produced from a heat source. The higher the temperature of the source, the more efficient the process is. A number of times throughout this book, I discuss higher temperature operation as a way to increase efficiency.

A large power plant with a conversion efficiency of 33% may generate 3,000 megawatts thermal (MWth) of heat energy but only 1,000 megawatts electric (MWe) of useful electricity. Throughout this book, always remember that the small "e" following MWe means electrical output, and a small "th" following MWth means thermal output. A 1,000 MWe (or 1 GWe) power plant can provide electricity for about 750,000 homes.

Units of electricity are units of energy and are typically given as kilowatt-hours (kWh). The amount of electricity produced by a 1 kWe power plant for 1 hour is 1 kWh—this is the amount of electricity required to run a 100-watt light bulb for 10 hours. The units of electricity can also be scaled up as megawatt-hours (MWh) or gigawatt-hours (GWh).

Climate Change

Scientists from around the world have warned that we need to take drastic measures now to prevent catastrophic problems caused by climate change in the future. Even if our country pushes aggressively now for reduced emissions, it will take many decades to slow climate change and on the order of centuries to reach a stabilization point.[3] Preindustrial carbon dioxide (CO_2) levels in the atmosphere were 280 parts per million (ppm), while current levels are at 375 ppm. To keep CO_2 levels below 500 ppm will require capping current carbon emissions for the next 50 years.[4]

Our country releases about 5.4 billion metric tons of CO_2 into the atmosphere each year, or about 17 metric tons per person per year (17 metric tons is equal to 17,000 kg). Roughly 43% of our emissions are from oil, 35% are from coal, and 22% are from natural gas.[1]

CO_2 emissions are not currently regulated. Most industrialized nations (except our country) have signed and ratified the Kyoto agreement, which is an attempt to limit greenhouse gas emissions to 1990 levels. Although this agreement is well meaning, the implementation could have far-reaching economic and even environmental consequences. Such consequences are described later in the book, but what is important is not to devastate the economy with expensive regulations. Our country is driven by market forces. Government regulations or subsidies can help to force change, but realistically, regulations will not garner support if they will force an industry out of economic competitiveness.

We need realistic development paths that allow our country to reduce emissions without spending trillions of dollars. Change does not occur overnight, but changes must start *now* so that in 30 years we can dramatically reduce emissions. New clean energy standards will need to be enacted gradually to give industry time to adapt and to have a chance of actually being passed as legislation.

Also, climate change is a global phenomenon, and in the future the United States may play only a small part in the entire world's greenhouse gas emissions. By 2035, world coal consumption is expected to increase by 56%. China alone is expected to add 736 new large coal power plants.[5] It is imperative that clean energy options become economically competitive without government subsidies to find acceptance throughout the world, as not all countries will have our same views on the environment.

I discuss life cycle emissions in this book when evaluating the different power options. This analysis takes into account every bit of pollution that is produced from constructing and operating a power plant to mining and transporting the fuel. The pollution during construction can include everything from the emissions generated while producing the materials (like steel and concrete), emissions from the electricity required, and even emissions from the vehicles used in construction. It also takes energy to extract, process, and transport fuel, and these energy requirements also produce emissions.

Conservation

Conservation and energy efficiency are two key focuses of the green movement today. While these are worthy goals and have helped to limit energy growth in the past, conservation by itself will not make a major impact on reducing current emissions. It is difficult to estimate how much conservation can help curb energy demand, but looking at historical data is useful. The idea of conservation has been around for a while; energy-efficient appliances are mainstream, and more efficient methods for home heating have been around for years. But in the past 20 years, energy demand has always increased from the year before. The reason is that the country's population is always growing, and that works against conservation.

Economics are a key reason why conservation has not made a major impact in the past. Energy-efficient appliances often cost more money than alternatives. More efficient ways to heat and cool a home also can require significant investments. The simple truth is that many Americans cannot afford to make these investments.

The only time that a region of the country was able to significantly reduce energy demand was following the blackouts in California in 2001. Due to a number of programs to get industries to reduce energy use and asking individuals to conserve energy, California was able to reduce energy consumption by 6.7%.[6] But that was only one state, and the entire country's energy demand still went up during that time. It is interesting that only in an energy crisis was an area of the country able to make an impact on energy demand through conservation, and even that impact was minor.

To truly tackle the problem of climate change, the country needs to focus on the dominant sources of emissions: burning oil for transportation, and burning fossil fuels for electricity. Although conservation and energy efficient technologies for the home can play a small role, these options are outside the scope of this book.

National Security

Ensuring a reliable domestic energy supply is critically important to national security. No matter what side of the fence a person is on politically, there is no doubt that one of the major reasons for our involvement in the Middle East is to try to increase stability in an area that contains one of the world's richest deposits of oil. The United States currently imports about 2/3 of the crude oil it uses.[1] It is in our country's best interest to move away from these foreign sources of energy.

Our involvement in the Middle East has been a source of political controversy for some time. The price tag from the Iraqi war in deaths and injuries has been very high, not to mention the billions of dollars spent on the war effort. How much of the billions of dollars spent on the war is justified by securing energy resources? And how much further along would energy research be if only a fraction of that money augmented current programs?

This argument goes well beyond just the United States. Reliable energy sources make the standard of living better in all countries. Reliable energy leads to cleaner water, less disease, better health care, and technological advance. Countries that have reliable energy sources tend to be more stable than those whose energy sources are extremely limited. All countries should be able to find the means to generate clean power self-sufficiently, and there is no reason why the United States cannot lead the world in developing these new technologies.

The domestic resource potential is a major driver for all the power options. The resources available in the country determine how prudent it is to continue relying on a particular energy option. When it comes to natural resources, I usually refer to fuel supply in this book in terms of economic reserves and total reserves. Economic reserves are an estimate of how much of that resource is available that can be extracted at current prices. Total reserves are an estimate of the total amount of that resource in the ground (whether economical to extract or not). These estimates are constantly changing as new sources appear or as technology improves to reduce extraction costs. Be careful when evaluating reserves; use these numbers only as a rough guideline.

Economics

In a capitalist society, economics is probably the most important driver for any new technology. Electricity has to be competitive for a particular power source to be even considered for large base load capacity. With all of the new energy technologies available today, it is interesting that most of our power is still produced by burning fuel. It is incredibly difficult for new energy options to compete with the economics of more established technologies. Coal power plants may seem archaic, but they have been around so long that the technology is highly optimized and the infrastructure is well established.

New energy technologies must have a clear development path headed toward widespread use. Such a development path may involve getting industry interested in stages or by government subsidies that help make the technology competitive. The purpose of government tax incentives or subsidies is to give new technologies the time to gain the experience needed to bring down costs to competitive levels. Gradually these incentives can be removed and competition can take over. But often it will only be through large-scale industrial experience that a new energy option will be able to be optimized enough to bring down costs to competitive levels.

Most cost comparisons of different power plant options usually look at the capital cost and the levelized cost of electricity. "Overnight" capital cost is typically used as a way to compare different energy options. This is the cost of building the power plant if it could be constructed overnight. Realistically, a plant may take a few years to build, and this adds costs such as interest during construction. Overnight capital cost is given in terms of $/kWe installed capacity. For example, if it would cost $2 billion to build a 1,000 MWe plant, the capital cost would be $2,000 per kWe installed.

The levelized cost of electricity is the baseline cost to generate the electricity in terms of ¢/kWh and is the most useful way to compare the costs of energy sources. It takes into account the capital cost, all plant financing, interest rates, tax, fuel costs, operation, and maintenance. This cost of electricity is not what the end consumer pays but rather the minimum cost for the utility to produce the power. After taking into account transmission costs and profits, actual electricity costs are often much higher. A competitive levelized cost of electricity may be around 6-8 ¢/kWh, but as consumers we typically pay around 8-10 ¢/kWh on our electric bill.

Acceptability

A major problem with expanding power production is the "not in my backyard" philosophy. Our country is critically dependent on electricity, yet there is always a great deal of resistance toward building new plants. We all want the conveniences of modern life, but without any of the negative consequences of the technology. We want to live in that nice home in the suburbs, but we do not want to build the new power plants to provide power for all the new subdivisions. Public acceptability of power plants has proven to become

The Energy Challenge

one of the most important factors in deciding whether a new plant gets built. Why is it that energy is so important to our world, yet so many of us take it for granted?

The first reason is that energy is often produced miles from our homes. In the past couple of centuries, energy sources have gradually moved away from the home toward centralized power stations. Imagine how much work it would be to still need to chop wood for the stove. Imagine how dirty it would be to have a coal furnace. Our modern homes are designed for heating and cooling with the flip of a switch, but it also removes us from caring about where the energy is coming from.

It is also easy to take energy for granted when our sources are extremely reliable. For the past twenty years oil prices have fluctuated, but gasoline has always been available. Electricity rarely goes out; but when it does, our dependence becomes apparent. Have you ever been in your house during a power outage, and you find yourself repeatedly walking into rooms flipping on the light switch? How amazing is it that something so important becomes second nature to us.

Just as life is not risk free, there are risks with every energy option. Accidents will occur with all power technologies. Government does its best to keep accidents from happening through regulations, but a little risk will always be present. Consumers inherently accept risk all the time when buying things like cars, airline tickets, a new stove, etc. It might be possible to build an airplane that could crash without injuring passengers, but each plane ticket would probably cost $50,000. Whether consumers realize it or not, they constantly trade value for risk, and that is okay for most—the key with energy technologies is to learn to put the risks in perspective.

Reliability

Electrical demand continuously cycles through periods of low and high demand on both daily and seasonal cycles. The ability of utilities to match power output with demand requires complex control systems. The country's electrical supply consists of many large base load plants, such as coal and nuclear, which produce a constant maximum supply of power. It is difficult to change the power output from these plants on a fast time scale, so often other energy sources (such as natural gas turbines) are used for peaking generators that can quickly come on line in times of high demand (like in the summer when air conditioners are running full blast).

As evident by the Northeast blackout in 2003, today's power grid is stressed. Reliable power plants with high capacity factors are needed to satisfy increasing demand. The "capacity factor" is typically used as a measure of reliability and is a measure of the average power available divided by the peak rated power of the plant. For example, if a large coal plant needs to be down for 15% of the year for maintenance or other reasons, it will have a capacity factor of 85%. In an increasingly competitive electricity market, utilities try to push the capacity factor as high as possible for cost effectiveness.

Reliability becomes a challenge with technologies like solar and wind that are intermittent. With a low average capacity factor due to the weather, it becomes difficult to depend on these renewable options for a significant fraction of power generation. A low reliability ultimately increases the cost of the system. If power sources cannot produce power constantly, backup systems will be required to make up for down times.

The Whole Picture

Achieving a brighter energy future is not just about one or two technologies. It will take a well-developed portfolio of clean power generation technologies along with an advanced transportation sector to reach our goals. I have broken this book down into four parts to examine each grouping of technologies in more detail.

Part I starts with the transportation sector with a focus on technologies that will get us away from oil. Part II examines each of the renewable energy options by taking a hard look at how feasible these options can be. Part III discusses nuclear energy, which is quite a bit different from all other sources of power. Finally, Part IV examines the fossil fuel generation technologies with an emphasis on eliminating carbon emissions.

The final chapter of this book presents an energy construct that summarizes the key technologies that need to be developed to solve our energy problems. We need a realistic energy plan to achieve a better energy future. Our government needs political leaders who can make these changes in the face of strong industrial lobbies. We need more support for research and development programs for clean energy technologies. We face an enormous challenge to achieve a clean, domestic, and economical energy future, but conquering this challenge will be one of the defining turning points of our society.

Part I

The Transportation Sector

AS THE LARGEST SOURCE of primary energy in the United States, oil quite literally drives the economy. There is no question that our country is addicted to oil and that the economy and our national security are strongly dependent on foreign sources of oil. In filling up at the pump, we are constantly faced with gas prices and world events that cause those prices to fluctuate. The use of oil is also responsible for a significant portion of our CO_2 emissions. But with our entire way of living so dependent on oil, what are the alternatives?

A significant amount of media attention has been devoted to the use of ethanol and hydrogen as possible alternative transportation fuels. However, both of these options have fatal flaws that I discuss in the next two chapters. The only transportation option that makes any real sense for achieving a clean, domestic, and economical energy future is the widespread use of electric vehicles.

Chapter 2: Biofuels

WITH A GREAT DEAL of publicity about biofuels, it makes sense to question how good these options are at reducing dependence on foreign oil and pollution. The term "biofuel" gives the impression of a green technology, and proponents argue that it leads to a sustainable transportation fuel. But this chapter describes three fatal flaws with biofuel use that must be taken into account when considering an alternative transportation future.

The most common biofuel in the United States is the production of ethanol from corn. In 2009, corn ethanol displaced about 3% of all gasoline sold in the United States.[1] Ethanol producers currently enjoy a $0.45 tax credit per gallon, but legislation may eliminate that credit in the future. The corn subsidy in our country is very high and totaled about $42 billion from 1995 to 2004.[2] The total subsidy, including all factors for ethanol, is estimated at about $0.89 per gallon.[2] One wonders if these subsidies are worth it or if this money would be better spent on other causes.

Production of Biofuels

Ethanol

Ethanol is created from food crops such as corn, soybeans, or sugarcane—the sugars in the food are converted to ethanol, which is a pure alcohol. The basic process is really no different than illegal production of moonshine, but the production rate is ramped up to the industrial scale. A combination of enzymes and yeast are added to the feedstock and heated—the resultant ethanol is purified for use as a transportation fuel. CO_2 is released as a by-product of this process, which needs to be considered when examining the environmental impact.

In the United States, ethanol is mixed with gasoline to increase the fuel octane rating and decrease some of the more harmful carbon monoxide emissions. However, the amount of CO_2 released by burning ethanol is the same as with gasoline. Any car can burn E10 which is a 10% ethanol - 90% gasoline blend. Specially designed flexible fuel vehicles can burn E85 which is an 85% ethanol - 15% gasoline blend.

Because corn is a dominant crop in the United States, a majority of ethanol produced domestically is from corn. It may be possible to also use cellulosic crops to produce ethanol, including wood, stalks, or grasses. However, the cellulosic conversion process is still in development and currently requires very high energy inputs.

Biodiesel

Whereas ethanol can replace gasoline, biodiesel can replace diesel fuel. The production of biodiesel is quite a bit different from ethanol and requires some type of fat or oil as the feedstock. Either animal fats or vegetable oils can be used. Algal biodiesel research has grown in recent years due to the potential to reduce energy and land requirements. Most biodiesel used in the United States is created from soybean oil or recycled cooking oils through various chemical reactions using an alcohol and catalyst solution.

An advantage of biodiesel is that most diesel engines can run using this fuel without any modifications. Biodiesel does not affect the performance of a vehicle much, though it contains slightly less energy than conventional diesel.

Net Energy Analysis

The first fatal flaw with the use of ethanol and other biofuels centers around the net energy analysis, which examines how much useful energy can be extracted from the fuel as compared to the energy required to produce it. Whether using corn to produce ethanol or using soybean oil to produce biodiesel, it takes a significant amount of energy to produce the fuels, so the question is whether the end result is a net source of energy.

Corn Ethanol

There has been some considerable debate about the net energy balance of corn ethanol. In reviewing the literature, it seems that most of the references written by those in support of ethanol show a slight energy gain (which is not surprising).[3] A number of other reviews show a net energy loss.[2] To try to get a better idea of the true answer, I examined the averages from some of the different studies.

Table 1 shows the average energy inputs that go into growing corn. The units on the right-most column are given as megajoule per liter (MJ/L), which represents the amount of energy that goes into producing one liter of ethanol. It takes energy to produce the fertilizers, herbicides, and insecticides required for producing good crops. Fossil fuels are used for farm machinery, transportation, and other uses. Energy also goes into the production of corn seed and for electricity. All of these inputs add up, but by far the largest input is the energy used for the actual conversion process in the refinery.

The total energy input is 23.3 MJ/L. But the actual energy content of ethanol is only about 22.3 MJ/L, so the production of ethanol results in a net energy *loss* of 1.0 MJ/L. There are variations in net energy studies due to the number of different assumptions that can be used in the calculation. Time could be spent debating that some numbers should go up or down slightly, but in general it is clear that ethanol production does not produce any net energy.

		Energy Input (MJ/L)
Field Chemicals	Nitrogen (Ammonium Nitrate)	2.73
	Phosphorus (P_2O_5)	0.18
	Potassium (K_2O)	0.22
	Calcinated Line (CaO)	0.22
	Herbicides	0.32
	Insecticides	0.04
Fossil Fuels (Machinery, Transportation)	Gasoline	0.49
	Diesel	1.12
	Liquified Petroleum Gas	0.45
	Natural Gas	0.40
Other Field Inputs	Corn Seed	0.79
	Electricity	0.43
Refinery	Ethanol Conversion	15.90
	Total Energy Input	**23.30**
	Energy Content of Ethanol	**22.30**
	Net Energy Gain	**-1.00**

Table 1: Energy Inputs to Grow Corn[2,4,5,6]

So where do all of these energy inputs come from? The production of field chemicals usually requires industrial heat sources from coal or natural gas. Farm machinery and transportation require oil. The ethanol conversion process requires heat, usually from natural gas. Therefore, most of these energy inputs are ultimately from fossil fuels, which means that the overall process is polluting.

Other Biofuels

Studies have also examined the net energy analysis of ethanol or biodiesel from other sources. Ethanol from sources like switchgrass or woody crops, and biodiesel from soybeans or sunflowers led to the same or worse results than ethanol production from corn.[7] Cellulosic ethanol from switchgrass or woody crops needs more research and development, but for now the process is still very energy intensive.

The one example of a positive net energy result for biofuels is the use of sugarcane for ethanol production in Brazil. Brazil has heavily developed their ethanol industry to reduce oil use over the past 20-30 years. The studies on this net energy analysis are limited, though results show an energy output that is 8-10 times the energy input.[8] There are a few reasons for this discrepancy from corn ethanol. First, sugarcane grows in a tropical, year-round climate. The sun's rays are intense near the equator and farm productivity is high. Second, the industry has become very efficient at ethanol conversion. Lastly, the sugarcane remains are used to produce most of the energy required for the conversion process, eliminating fossil fuel requirements for that step.

Unfortunately, the use of sugarcane is not available in the United States due to the more northern climate. Also, Brazil uses only about 4% of the amount of gasoline that the United States uses, so they are more easily able to use their land for this purpose.[9]

Environmental Impact

Net Emissions

The net emissions of ethanol from corn are directly linked to the net energy analysis and lead to the second fatal flaw of biofuel use. About 70-75% of the energy inputs required to produce ethanol from corn come from coal or natural gas, 4-5% come from other sources of electricity (like nuclear or renewables), and 5-20% come from oil.[10] Since the majority of the energy inputs come from fossil fuels, CO_2 and other pollutants will be released into the atmosphere.

Once again, studies vary as to the amount of CO_2 released from the use of ethanol as compared to gasoline. One study by the International Energy Agency found an average 20-40% reduction of greenhouse gases from ethanol.[6] Though, this study tended to leave out some energy inputs. Another reference showed that ethanol use will ultimately result in a 50% increase in greenhouse gas emissions as opposed to gasoline use.[5] Again, the true answer is likely somewhere in the middle, which suggests that ethanol use is just as polluting as gasoline.

In these analyses, the CO_2 release from the actual burning of the ethanol is ignored since it is assumed that to grow, the corn crop absorbs an equal amount of CO_2 from the atmosphere. The emissions are all from the fossil fuel energy inputs. The key point is that ethanol use will make no difference on greenhouse gas emissions as compared to using oil.

Land Requirements

The final flaw with biofuels is that even if the entire U.S. corn crop were converted to the production of ethanol, it would displace only about 15% of our oil use.[2] It takes up an incredibly huge amount of land to grow corn, so ethanol will not make a dramatic difference on foreign oil imports.

Corn is one of the most energy-intensive and unsustainable crops. It requires tremendous amounts of fertilizer, but even then depletes the soil substantially. Fertilizer runoff can have certain environmental consequences. The term "biofuels" tend to give us an impression of helping the environment. Ultimately, though, because of the CO_2 emissions and fertilizer runoff, the use of ethanol or other biofuels is only going to cause more environmental harm than the alternatives.

Global Effects of Biofuels

A past article in the *New York Times* examined an "Econightmare" that was the result of pushing for biofuels.[11] Politicians and environmental groups in the Netherlands pushed for the use of biofuels as a sustainable energy source from palm oil grown in Southeast Asia. The demand of palm oil increased as a result of government subsidies, and this pushed for more palm oil production. Areas in Southeast Asia cleared huge tracks of rainforest and peatland to expand the palm plantations, in some cases burning the forests down. The release of CO_2 into the atmosphere from this debacle completely defeated the purpose of encouraging biofuel use and led to a valuable lesson. In a global economy, we need be aware of the effects of so-called green subsidies on other parts of the world.

Conclusion

The three flaws of biofuels show why it makes little sense to develop these fuels as an alternative to oil. The key problem is that biofuels will be just as polluting as gasoline, so they will do nothing to help prevent climate change. Since the country simply does not have enough extra farmland to make a significant dent in the use of oil, biofuel use will do very little to reduce dependency on foreign sources of oil. Biofuels can make a small contribution by utilizing excess crops, but it is a waste of money and resources to continue subsidizing biofuels since they will never be able to make a major impact on pollution and oil use.

In the future, it would be a much better use of farmland to produce food for an expanding population or to restore forests to help absorb excess CO_2 in the atmosphere. Fortunately, alternative transportation options exist that will allow the country to make a more dramatic impact on oil consumption while at the same time reducing pollution.

Chapter 3: Hydrogen

IN RECENT YEARS hydrogen has received a great deal of attention as a transportation fuel of the future. The idea behind the hydrogen economy is that cars burn hydrogen and oxygen to produce power, and the only by-product is water. In the 2003 State of the Union Address, President George W. Bush announced a major hydrogen initiative, proposing $1.2 billion for hydrogen and fuel cell-related research. But like biofuels, the hydrogen economy has flaws.

First and foremost, hydrogen is an energy carrier, not an energy source—it must be created in a power plant. There will be a great deal of energy loss in the entire process from the creation of hydrogen, to distribution, to use in fuel cells. The primary energy requirements for a hydrogen economy will be enormous, almost twice as much as if ethanol or electric vehicles were used instead. Furthermore, hydrogen must be created in a clean power plant using nuclear or renewable technologies in order to have any environmental gain.

Second, an entirely new distribution infrastructure will have to be created to use hydrogen as a transportation fuel. Many technological challenges exist, such as whether it makes more sense to distribute hydrogen as a gas, liquid, or in solid forms. Along these lines, hydrogen storage in the vehicles will be challenging. Even if these problems can be solved, there will still be a huge cost associated with making this transition to a new infrastructure.

Lastly, fuel cells that create energy from hydrogen are currently expensive and not optimal for a vehicle's power source. There is a tremendous amount of research required to choose the most optimal design and bring down costs. Hydrogen has some tough competition from hybrid vehicles, which are achieving much better efficiencies than the traditional internal combustion engine. A number of experts believe that it is highly unlikely that hydrogen-powered vehicles will be economical and available in the next 30 years.[1,2]

Hydrogen Technology

The proposed hydrogen economy would be based upon a huge number of power plants to produce the hydrogen and a completely new distribution system. Instead of heading to the gas station to fill up your tank, you would instead head to a hydrogen station. The cars themselves would run on electric motors powered by a hydrogen fuel cell. The following sections describe the key technological components of the hydrogen economy.

Creating Hydrogen

Hydrogen currently is produced from fossil fuels through steam reforming, which combines steam and hydrocarbons to produce hydrogen and carbon dioxide. However, since this process produces a greenhouse gas and uses up fossil fuels, it defeats the purpose of moving toward a clean transportation option.

To have a clean source of hydrogen, it must be produced from water using clean sources of power. One method, called electrolysis, involves the splitting of water to produce hydrogen. (Water is composed of two hydrogen atoms and one oxygen atom, but it requires energy to remove the hydrogen from the oxygen.) If electrolysis were used, the energy efficiency for producing hydrogen would be around 30%.[3]

A more efficient process for creating hydrogen is the sulfur-iodide cycle.[4] This cycle is a chemical process that splits water at very high temperatures (around 800-900 °C) using the heat from a power plant. This cycle could create hydrogen with an efficiency near 45% or higher; however, a great deal of research will be required for this to be commercially viable.

Hydrogen Storage

Once hydrogen is created at the power plant, it must be distributed across the country and safely stored within vehicles—leading to another engineering issue for the hydrogen economy. Hydrogen can be transported and then stored in a car either as a compressed gas, a cryogenic liquid, or in a solid metal hydride. But even in its most dense form, hydrogen takes up much more volume than gasoline for the same energy content.

Hydrogen can be compressed for transportation, but realistically, the maximum pressure achievable is about 700 times atmospheric pressure, for which the energy content per volume will be 5 times lower than an equivalent volume of gasoline.[5] In addition, it takes energy to compress the gas to these high pressures, which results in a loss of energy. The other challenge with compressed hydrogen is in developing a tank to contain it safely that does not weigh too much. Carbon composites are being examined for their high strength and low weight.

The second distribution option is to cool hydrogen until it condenses out as a cryogenic liquid, which requires reaching temperatures of -259.2 °C. Liquid hydrogen results in about the same volume requirements as very highly compressed gas; however, a heavily insulated tank will be required to withstand the pressure created by boil off.

Hydrogen could also be stored in metals as metal hydrides. Because hydrogen is such a small molecule, it can migrate into and be stored within the space between metal atoms. The storage volume is similar to highly compressed hydrogen or liquid hydrogen but without the pressure. However, metal hydrides will require much more research and development for storage and fueling.

Hydrogen Fuel Cells

Using hydrogen stored in the car plus oxygen from air, a hydrogen vehicle generates its energy from the combustion of the two elements. A fuel cell is an efficient process for directly creating an electric current from this combustion process that can then be used to power electric motors.

Figure 2 shows a general schematic of a proton exchange membrane (PEM) fuel cell. Both hydrogen and oxygen are supplied to different sides of the fuel cell. A platinum catalyst removes the electron from hydrogen, leaving only a proton. A proton exchange membrane separates the two cathodes and allows only protons to pass through—this in turn produces an electric current. The only by-product of the reaction is water, which is released as water vapor.

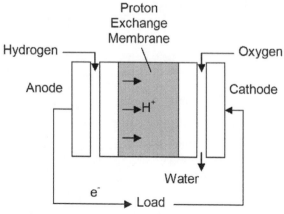

Figure 2: Hydrogen Fuel Cell[5]

Hydrogen fuel cells work much like a battery, except the fuel is constantly supplied. In a car, the electric current from the fuel cell directly powers an electric motor. Fuel cells currently cost about $4,000 per kW, which is about 100 times greater than the cost of an internal combustion engine.[1,2]

Proton exchange membrane (PEM) fuel cells, like that shown in Figure 2, appear to have the most promise for use in automobiles. Although some references suggest that fuel cells could reach efficiencies from 60-70%, these efficiencies are theoretical and not based on real designs. The hydrogen to electricity efficiency currently achieved for PEM fuel cells is around 35%, but it is hoped that efficiencies closer to 50% can be reached in a real design.[5] The efficiency for electrical production in a fuel cell is important in understanding one of the key problems with the use of hydrogen, as described in the next section.

Net Energy Analysis

For a while, investigating hydrogen production was the latest fad in energy research. A number of studies have evaluated everything from using wind and solar power to nuclear power for producing hydrogen. As I started seeing more of these studies come out, I could not help but wonder how efficient the overall process could be. While doing the research for this book, I realized a key problem that had been overlooked with the hydrogen economy.

The net energy analysis of hydrogen shows that there are a number of efficiency losses in the overall process of creating hydrogen, transporting it, and finally using it in a fuel cell. In the best-case scenario, it can be assumed that hydrogen will be created with the high temperature sulfur-iodide cycle with a production efficiency of about 45%. Currently the most energy-efficient process to distribute

Hydrogen

hydrogen is to compress it. Compression has an efficiency of 90%, and transporting the compressed gas has an efficiency of 80%.[3] In other words, it takes energy to compress and transport the fuel, which will result in losses. Once the hydrogen is in the car, the fuel cell at best may have an efficiency of 50%, though this could be stretching reality. Finally, the vehicle's use of the fuel cell current to power the electric motor may have an efficiency of 90%.[3]

Multiplying all of these best-case efficiencies, the overall hydrogen system efficiency is:

$$0.45 \times 0.90 \times 0.80 \times 0.50 \times 0.9 = 0.15 \ (15\%)$$

The useful energy output from hydrogen will be only about 15% of the amount of primary energy produced in the power plant—meaning about 85% of the primary energy is wasted. This is a huge loss of energy, and it will be incredibly wasteful of energy resources!

The power options that would produce the hydrogen also need to be considered. The use of coal or natural gas plants does not make sense because they would generate emissions. Renewable energy options like wind have been examined for producing hydrogen, but the hydrogen would need to be formed using electrolysis which is less efficient. The only other option is to use advanced high-temperature nuclear reactors to produce all of the hydrogen. At this overall efficiency, about 860 new large 1,000 MWe nuclear reactors would need to be added immediately in order to produce enough hydrogen to fuel the entire transportation sector.

Note that these efficiency estimates, if anything, are optimistic, so the actual energy losses could be even greater. Even at this optimistic level, the hydrogen economy would waste twice as much primary energy as the use of electric vehicles, which is described in the following chapter.

Conclusion

Above all other reasons, the low overall efficiency of the hydrogen economy is the greatest downfall of this idea. Hydrogen is simply too wasteful of energy resources, which will ultimately lead to more pollution. On top of that, the required major infrastructure changes and lengthy research time make hydrogen technologies impractical. We need alternative transportation options that are available now, not options that *might* be available in 30 years. All in all, the hydrogen economy does not make sense. Our country should not be wasting limited resources on further research of these technologies.

Chapter 4: Electric Vehicles

SINCE BIOFUELS AND HYDROGEN make little sense as a clean transportation alternative, only one other option remains: the electric car. A number of years ago, I never would have supported electric vehicles. The electric cars of the past have been limited in size and range, and the fact that they need to be plugged in for long periods of time limits their use. I am the kind of person who wants a car that can do everything. I always look for a car with good gas mileage but one that also has power. I want a car that is good for road trips, with enough room for luggage or camping gear. The electric car of the past did not fit my needs.

But that was the electric car of 20 years ago. As I starting researching this area, I soon realized that instead I needed to be thinking about the electric car of the future. Imagine a full-sized electric car with a range of 300 miles and the power of a V6. Imagine an electric motor and battery pack that take up the same amount of room as all the components of today's internal combustion engine. Finally, imagine a car that can be rapidly charged in a few minutes. These technologies may not be too far off.

Compared to ethanol and hydrogen, the use of electric vehicles has a number of advantages. New infrastructure does not need to be developed since electricity is already available everywhere. Electric vehicles will require the addition of a large number of new power plants, but half as many new plants would be required as compared to the use of hydrogen. Also, the commercial success of hybrid vehicles presents an excellent development path toward reaching the all-electric car—the electric engine and battery components are already being installed in hybrids. From the standpoints of economics, the environment, time of development, and infrastructure requirements, electric vehicles *make the most sense*. Fortunately under the Obama administration, electric vehicles have started to receive the political support that they deserve.

Electric Vehicle Technology

Electric vehicles use an electric motor that runs on a battery. Due to their quiet and smooth operation, zero emissions, and high efficiency, electric motors have many advantages over the internal combustion engine. The engineering challenges of electric vehicles are all associated with the battery technology including developing compact batteries, extending the range, increasing battery life, driving down costs, and moving toward rapid-charge capability.

In an electric vehicle, the pedal and brake regulate the power flow to the motor. Efficiencies around 90% can be achieved over a wide range of speeds and power outputs. Unlike the internal combustion engine, electric motors do not require gear boxes to increase or decrease torque. Electric motors can deliver whatever power requirements are desired for a car, but power has been limited in the past to keep the battery size down. As battery technology progresses, the size and power of electric vehicles will gradually start competing with full-sized cars and trucks.

Battery Technology

It may be difficult to imagine batteries powering the entire transportation fleet. It is one thing to provide power for a compact car, but what about a full-sized truck? But consider

Electric Vehicles 35

how far battery-operated power tools have come in the past couple of decades. There was a time when a general contractor would not be caught dead with a battery-operated drill. Yet now, these types of tools pack a lot of power.

The batteries used for electric cars have always been the largest technological hurdle. Batteries take up space and add extra weight to a vehicle. The cost of the batteries is also high. The long charging time of electric vehicles in the past made them undesirable for many people. But all of these issues can be addressed with continued research and development.

There are a number of battery types that can be useful for the electric vehicle application: lead-acid, nickel-iron, nickel-cadmium, nickel metal hydride (NiMH), lithium polymer, and lithium ion batteries. The key is to develop a battery with a high energy density to minimize the space needed in the vehicle. At the same time, the battery must have a long life and be economical.

Lead acid batteries have been used for automotive batteries for a long time, so their cost is low and the technology is mature. But they have a relatively low energy density, and they have safety issues due to the presence of toxic materials. The NiMH battery has the highest specific energy and specific power of the nickel-based options. It is more environmentally friendly, has good discharge characteristics, and is capable of being charged rapidly. From 1997 to 2004, the price and weight of NiMH batteries used in hybrids dropped by half. However, the batteries used in hybrids still account for about 50% of the extra cost.[1]

Lithium batteries probably have the most potential for powering the electric car of the future, as they have the highest power density. The large-scale commercial use of lithium batteries in cell phones and laptops has allowed the technology to be developed and has driven down costs in recent years. Newer designs are very safe, can be recharged in minutes, hold their charge longer, and can be used in a wide temperature range.[2]

For large-scale use of batteries in electric vehicles, the resource base of the required materials will make an impact. The United States has limited supplies of lithium, so other compact battery designs may be required. For this reason, it will be advantageous to maintain a broad research program that investigates a number of different battery options in parallel.

Rapid Charge Capability

There has been some limited work in developing rapid charge capability as a way to make electric cars more desirable to the average person. The Hawaii Electric Vehicle Demonstration Program has demonstrated rapid charge for a number of years with electric vehicles on the island of Oahu. A fleet of Hyundai Santa Fe electric vehicles were equipped with NiMH batteries and electric motors. Rapid charging stations around the island were able to charge the batteries in about a half an hour (their range was about 100 miles).[3] This range and charge time is still undesirable for most people, but it is a huge step in the right direction.

More promising is recent work by Toshiba and Altair Nanotechnologies to develop high-power lithium ion batteries for use with cars.[4,5] Toshiba has announced a lithium ion battery with a robust temperature range of operation, a high cycle life, and the ability to be charged to 80% capacity in about *1 minute*. If it can be manufactured economically, this battery could have a dramatic impact for electric vehicles.

Net Energy Analysis

To compare electric vehicles to hydrogen, it is also useful to determine the overall efficiency. The production,

Electric Vehicles

transmission, and use of electricity for electric vehicles also have a number of losses associated with them, but the overall efficiency is much better as compared to the hydrogen economy.

To keep the analysis consistent with the hydrogen analysis, a power plant conversion efficiency of 45% is assumed. Transmission losses in the grid are typically about 10%, so the transmission efficiency is 90%. There are also some losses in storing electricity in a lithium ion battery, which leads to an efficiency of around 90%. Finally, the motor efficiency within an electric vehicle is around 90%.[6]

The energy inputs that go into making the battery need to be taken into account as well. Depending on the battery type, 1.2-1.7 GJth of energy may go into manufacturing a battery with a 1 kWh capacity.[7] Assuming a battery of this size has a lifetime of 2,500 cycles, it will be able to deliver a total of 2,500 kWh worth of electricity to a car before needing replacement. And 2,500 kWh of electricity uses up about 20 GJth of primary energy, so the energy input to make the battery is only a small percentage of the energy needed to power an electric vehicle. In other words, there is another loss of roughly 10% due to the energy requirements for manufacturing the batteries.

Then the final net energy efficiency of the overall system will be:

$$0.45 \times 0.9 \times 0.9 \times 0.9 \times 0.9 = 0.30 \ (30\%)$$

Therefore, the overall efficiency using electric vehicles is about 30%, or twice the overall efficiency of the hydrogen economy. In other words, by using electric cars instead of hydrogen fuel cells as the transportation of the future, the country will need only to build half as many new power plants. Note that clean energy technologies would be required in order for the use of electric vehicles to have an environmental gain, but any clean producer of electricity (from zero emission coal to nuclear to renewable energy) can achieve this goal.

How much new electrical generation will be required? The transportation sector energy use for 2009 was about 29 billion gigajoules (GJ), and about 94% of that was from oil.[8] If this entire sector were replaced with electric vehicles, about 410 large 1,000 MWe power plants would need to be added to satisfy the energy demand. This is an enormous number of new plants to add, but they would be domestic energy sources.

Development Path

The commercial success of hybrid vehicles is one of the major reasons why electric vehicles will be the clean transportation of the future. Annual sales of hybrids were 270,000 in 2010 after a couple years of stagnation due to the recession. Hybrid gasoline-electric cars have probably done better than any auto manufacturer could have expected. They have shown that many in our country are willing to pay more for an environmentally friendly vehicle. With most hybrids, as long as the owner holds on to the car for 7 years or so, the added cost will be made up for by the better gas mileage.

Hybrid Technology

Hybrid vehicles use both an internal combustion engine and electric motor to increase the overall fuel economy. Since the electric motor can provide additional power, hybrids can get by with a smaller gasoline engine. This combination draws on the strengths of both technologies. Regenerative braking captures the energy lost in braking and stores it in a small battery. The configuration for the power train can vary; in some cases the gasoline engine and electric motor directly power the car, but in other configurations the gasoline engine acts as a generator to produce power for the electric motor.

Electric Vehicles

Full hybrids can use many other techniques to save energy throughout the car. Electrical components such as air-conditioning, power steering, and pumps can be made more efficient if the power is drawn directly from the battery. The engine can shut down when the car is stopped to save energy. Some hybrids switch engine types to more efficient cycles.

Hybrids have an advantage over traditional autos mainly because of the regenerative braking, so the fuel efficiency is much better for stop-and-go city driving conditions. The efficiency gains are not as good in comparing highway travel. Hybrids cost more because of the requirement for two power systems instead of just one. In addition, since hybrids are still young in terms of commercial maturity, their costs will be higher until the industry has more time to optimize the technology.

Plug-in Hybrid Technology

The next step in hybrid development will be the plug-in hybrid. The plug-in hybrid uses a larger battery that holds enough energy to drive a car for 20-30 miles (a typical commute for most Americans). In comparison, the factory-built Prius can run on battery for only about one mile.[9] With a plug-in hybrid, the battery can be plugged in at night for normal small commutes. However, it still has the gasoline engine and tank if a person forgets to plug it in, or for longer trips when plugging in is not feasible. The challenge for plug-in hybrids is that the battery pack will be expensive and add further cost to an already expensive car. Continued development and mass production will likely bring the costs down.

A number of hybrid enthusiasts have already made modifications in their garages to turn a hybrid into a plug-in hybrid. By swapping out the battery for a larger one and adding a plug, it is actually a fairly simple modification for a person with the right background. A person trying to modify his or her hybrid may have to spend $10,000-$15,000 for the battery pack.[9,10] However it is likely that a car company with the resources of Toyota, Honda, or General Motors could mass produce such a battery for considerably less.

The plug-in hybrid is an excellent development path toward achieving an all-electric vehicle. Only through industrial-scale development will the costs of the batteries come down and the charging technology be optimized to improve performance. Likely the first plug-in hybrid will have a battery range of 20 miles, but eventually competition will start extending that range. Then, once the rapid-charge technology is feasible, the gasoline component can be eliminated altogether.

The past year has seen the introduction of two new electric vehicles and more expected in the future. The Chevy Volt and the Nissan Leaf were recently introduced to the market. Both are producing only small volumes initially despite large waiting lists. The Leaf is an all-electric vehicle with a range of about 100 miles,[11] while the Volt is a plug-in hybrid with a battery range of 35 miles and up to 340 miles on a full tank of gas.[12] Continued development of these and other electric vehicles will help to drive down costs.

Economics of Electric Cars

A huge advantage of the electric vehicle is the cost savings from using electricity instead of gasoline. With gasoline costing around $3 per gallon, it costs roughly 12¢ per mile to fuel a car that gets 25 mpg. To power a car using electricity will only cost about 3¢ per mile[1]—a factor of 4 reduction in costs!

Another way to look at it is that the typical American may drive about 15,000 miles per year. Assuming 25 mpg and $3 per gallon gasoline, the typical person will spend $1,800 per

Electric Vehicles

year on gasoline. If an electric car were used instead, the typical American would spend only $450 per year on additional electricity costs (saving $1,350 per year). Even if an electric car costs $5,000 more than the gasoline equivalent, after about 4 years the fuel savings would pay for it.

Conclusion

Electric vehicles are the only advanced transportation option that will reduce emissions and use our domestic energy sources efficiently. The hybrid and plug-in hybrid technologies give auto manufacturers a clear development path toward reaching these types of vehicles in the future. But electric vehicles will make sense only if we move toward clean energy generation, the subject of the rest of the book.

Political support of electric vehicles has grown in recent years as the misconceptions about biofuels and hydrogen have been cleared up. The hybrids and plug-in hybrids will be the transition toward wide-spread use of all-electric vehicles that can occur gradually to allow time for the technology to optimize itself. If rapid charge technology is developed in parallel, it may be possible to have an all-electric vehicle that is desirable to a majority of the population in 20 years.

It may seem like a huge cost to have to build so many power plants over the next few decades to power electric vehicles, but the alternatives need to be considered. Continued reliance on oil will only keep oil prices on the rise and continue our dependence on unstable regions of the world. The time is well past to push strongly for an energy-independent and clean-energy future; electric vehicles will help us to reach that future.

Part II

Renewable Energy

THE ADDITION OF PLUG-IN hybrids to an already increasing electrical demand will make it that much more important to develop clean energy technologies to reduce CO_2 emissions. Renewable energy sources will be part of the solution of satisfying demand in a clean and sustainable manner. The next five chapters investigate solar, geothermal, biomass, hydro, and wind energy.

The development of renewable energy technologies has been accelerated through state renewable energy portfolio standards. These standards are state mandates that require utilities to fulfill energy demand with a certain percentage of renewable energy sources. In many cases, these mandates, along with government subsidies or tax incentives, have been the only way for some of the renewable options to expand.

Many of the renewable technologies still have an economic disadvantage as compared to other energy sources. As such, it may be a while before some of the renewable options see widespread use. On the other hand, a number of smaller-scale uses of renewable energy are being built today, and this portfolio could soon add up to a sizeable fraction of our power generation. Renewable energy technologies are some of the most fascinating ways to extract energy and are worthy of strong research support. The following chapters identify the technologies most likely to contribute to satisfying electrical demand in the near term and most promising for development in the long term.

Chapter 5: Solar Energy

WHEN I FIRST MOVED to New Mexico, I noticed that a number of homes in the area had solar panels installed on the roof. With blue skies and few cloudy days, New Mexico is one of the ideal locations in the country to take advantage of solar energy. Yet I also realized that most of the solar panels I saw were installed on older homes—I did not seen them on new homes.

One of the key problems with solar energy, which I will explain in this chapter, is the high cost of this option. A number of novel concepts for turning solar radiation into useful electricity have been demonstrated over the years, but the costs are far from competitive with other energy options. Installing solar panels on a new home may easily have an upfront cost of $50,000—this is simply too much money for builders who are trying to keep homes affordable.

The amount of solar radiation reaching the surface of the earth is staggering, which is why solar energy is too valuable an energy source to ignore. But the sun does not shine 24 hours a day, and unfortunately power plants are usually not economical when they cannot produce power constantly. There have been a number of large-scale demonstrations of solar plants in the past, but the economics never reached a point of attractiveness to commercial utilities. Even with strong government subsidies and tax credits, solar plants may be suitable only in areas like the sunny Southwest to provide competitive peaking power. Still, the solar-generating technologies are unique and one day may lead to a more economical plant.

Solar Technology

There are numerous concepts for a solar plant design, but four key methods stand out as having the most potential. Three of the methods use concentrated solar power to heat up a fluid, which then drives some type of thermal cycle. These three methods are the solar trough, the solar tower, and the solar dish collector. The fourth method uses photovoltaic cells to directly convert sunlight into electricity.

Solar Trough

The solar trough concept uses a series of reflective, parabolic-shaped mirrors that focus sunlight onto a tube running through the focal line. The focused solar power heats up a fluid traveling through the tube. To maximize the amount of energy absorbed, the trough can be rotated on one axis to track the sunlight as the sun moves through the sky. Oil is typically used as the heat-absorbing liquid, and may reach temperatures close to 400 °C. A simplified diagram of a solar trough power plant is shown in Figure 3.

Figure 3 shows how the oil is pumped through the system. As the oil passes through the parabolic reflectors, its temperature increases. Energy storage can be incorporated into the solar trough by creating a well-insulated tank into which the heated oil is pumped. With proper design, this tank can be large enough to contain enough hot oil to produce power through the nighttime hours. The high-temperature oil can be used to generate steam to drive a turbine to produce electricity.

The power range for a solar trough plant could be between 30-150 MWe,[1] making it a good candidate for large-

Solar Energy

scale power production. But due to the low reliability, the average annual solar to electric efficiency is about 11-15% with peaks at 23%.[1,2] This significantly hurts the economics of solar energy.

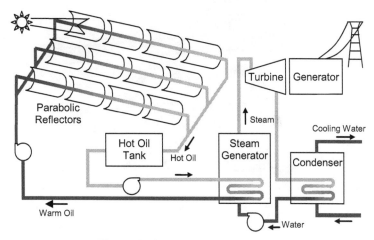

Figure 3: Solar Trough Concept

Solar troughs have the most commercial experience of the solar thermal options. Nine solar trough power plants were built in Southern California from 1984-1991 (see Figure 4). They are still producing a total of 354 MWe, but the technology was too costly to be able to justify building more systems. A 1 MWe solar trough power plant was just recently constructed between Phoenix and Tucson by Arizona Public Service Company, and a 280 MWe plant is currently being planned. A 64 MWe plant was recently built in Boulder City, Nevada. These recent efforts are encouraging but possible only because of state renewable energy portfolio standards or federal loan guarantees. However, such standards could help push for the industrial experience that solar power needs to bring down costs.

Figure 4: Solar Trough Power Plant (Courtesy of NREL)[3]

Solar Tower

In the solar tower concept, a field of mirrors called heliostats track and reflect sunlight onto a central tower. The movie *Sahara*, based on Clive Cussler's novel, shows a nice rendition of what a solar tower may look like. The concentrated light at the tower is used to heat a molten salt to high temperatures for use in driving a power cycle. The heliostats track the sun on two axes for maximum efficiency. The solar tower can also use energy storage by pumping the hot molten salt to a well-insulated storage tank. Figure 5 shows a schematic.

The power range for a solar tower can be between 30-160 MWe,[1] making this concept another option for producing large-scale power. Pilot plants have operated with molten salt temperatures around 550 °C, but it is possible to raise the temperature to 1,000 °C to increase overall efficiency.[2] Higher temperatures will require the use of ceramic materials in the

Solar Energy

receiver. The solar tower has never been proven commercially, and current research suggests that it may not be able to reduce costs as quick as other concepts.

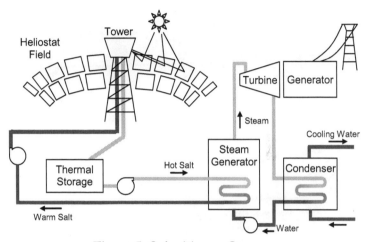

Figure 5: Solar Tower Concept

A number of solar tower pilot plants were built around the world in the early 1980s ranging in size from 0.5 to 10 MWe.[4] Coolants that have been used include water, sodium, and salts, but the molten salts have proven to be the most successful. The largest projects were Solar One and Solar Two in Barstow, California (both at 10 MWe). Solar One was operated from 1982 to 1988, and Solar Two (see Figure 6) was operated from 1996 to 1999. Although both projects provided valuable research on this technology, it was found to be too expensive to compete with other forms of power.

One of the issues to be careful with in considering alternative energy sources is the promise of using cheaper materials or higher temperatures to make the energy source more efficient, and thus more competitive. For example, the solar tower concept may be more cost-effective if it operates at a higher coolant temperature. However, operating at a higher temperature requires high-temperature materials,

Figure 6: Solar Tower Power Plant (Courtesy of NREL)[3]

which requires research and development. By the time these materials are developed, they can also be applied to more conventional power plants, like natural gas, coal, and nuclear, which would make them cheaper too. In the 20 years or so that an alternative energy source is making advances, the traditional sources are also becoming more efficient and streamlined. This fact makes it incredibly difficult for new energy technologies like solar to break into the market.

Solar-Stirling Collectors

Another solar thermal option is the solar-Stirling collector. This system uses a dish solar collector to focus the sun's power onto a small receiver. The receiver uses the heat to drive a Stirling engine, which is a different method for generating electricity from a heat source. Figure 7 shows a picture of the collector. The Stirling engine contains a working gas like helium to drive the heat cycle.

The individual concept is small in scale, typically about 10-50 kWe or so.[1] Therefore, large base load power would require many units in a modular design. The efficiency when the sun is shining has been demonstrated at 30% with the hopes of achieving 40%.[2] However, the average efficiency is just as limited as the other solar thermal options. This technology is still relatively expensive, and probably cannot

Solar Energy

compete with the solar trough or tower in the near term. It may find use for smaller applications though.

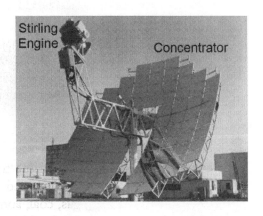

Figure 7: Solar-Stirling Collector (Courtesy of NREL)[3]

The solar-Stirling collector requires direct sunlight, so it must track the sun on two axes, which leads to high costs. Due to the nature of the collector, energy storage could only be done with battery backup. Due to the current high costs of battery backup, it may be more useful to use this solar technology just for power augmentation as opposed to base load power.

Photovoltaic Cells

Photovoltaic (PV) cells are semiconductors that directly convert sunlight into an electric current. Cells can be added in series or in parallel to create just about any voltage and current desired. PV cells operate with no moving parts, produce no emissions while in operation, and require little or no maintenance. They can operate using direct or diffuse light, so they are still operational in cloudy weather. The major drawback is the high cost of production.

The cost of PV systems has come down considerably in recent years as more efficient manufacturing processes have been developed. Whereas just a few years ago the solar thermal options appeared to have the upper hand, now PV is more popular with utilities since the installation costs have dropped quicker.

Silicon is typically used for PV because it is abundant and cheap to work with. There are a number of different types of silicon cells: single crystal, polycrystalline, amorphous, and thin-film. In addition, there are more exotic materials that have been studied. There is no clear winner amongst the different choices as usually the technologies that offer the better efficiencies are also more costly to produce. The most commonly produced single crystal and polycrystalline silicon PV cells have actual efficiencies between 8-12 %.[4] Higher efficiencies have been demonstrated—but usually at a much higher production cost.

One other alternative to solar PV is the use of concentrators to minimize the amount of PV area required. A less-expensive concentrator can be used to focus sunlight from a large area onto a relatively small PV cell. Cells with higher efficiency (which are more costly) can then be used. Unfortunately, concentrated PV does not seem to have any cost advantage over traditional flat plate PV at this point.[4]

Reliability

The inherent weakness in solar power is the low average availability of sunlight. The only way that solar energy can contribute to base load power is if it is coupled to an energy storage system or uses a conventional fuel backup. Energy storage systems are expensive and add to an already expensive technology; however, they are required for maintaining constant production of solar power at night and in times of cloudy weather. Even in the Mojave Desert, which has one of the most optimal conditions for solar in the United States, the average availability of solar power is about 25-30% (averaged over an entire year).[5]

Solar research in the past has led to a number of novel approaches to increase the availability. Hybrid solar-fossil fuel plants have been conceived to increase the reliability and decrease the cost. This book is focusing on investigating each energy option individually, so hybrid designs are not being considered. In reality, a utility that adds solar will need to maintain natural gas turbines that can come on-line during periods of low availability. And solar can match well with increases in summer electrical demand for cooling.

Environmental Impact

The net emissions analysis for solar power depends on the type of technology. The energy inputs include all plant construction costs, fabrication of mirrors or photovoltaic cells, operation and maintenance, and decommissioning. Although the solar options do not emit any emissions during operation, emissions will be released indirectly due to the energy requirements in manufacturing the components and during construction.

Net emissions analyses of photovoltaic systems show that this technology will release about 39 g of CO_2 per kWh.[6] An analysis of the solar tower finds an average release of 30-43 g of CO_2 per kWh.[2,7] (In comparison to coal power plants that can emit close to 1,000 g CO_2 per kWh, this is very small.) The production of photovoltaic cells can use toxic materials depending on the type of cell, but with proper planning it is relatively easy to prevent release of these materials into the environment.

Economics

Using a number of sources, the ranges of capital costs for the different solar technologies are shown in Table 2. The costs shown are for optimal locations in the country that receive plenty of sun such as a site in the Southwest. The following text goes into more detail about where the data came from to give an idea of the confidence in the numbers.

	Solar Trough	Solar Tower	Solar-Stirling	Solar PV
Capital Cost ($/kWe)	4,700-5,010	5,140-9,090	4,000-6,800	4,750-6,050

Table 2: Solar Power Economics

So how does this table compare to other energy options? All of the economic data presented throughout this book is compared in the Appendix. The most recent cost estimate projects the levelized cost of electricity at 31 ¢/kWh for solar thermal and 21 ¢/kWh for solar PV.[8] These numbers are high due to the low capacity factor of solar plants, but PV technology has become more competitive in recent years.

The capital cost for the solar trough with thermal storage has been estimated between $4,700-$5,010 per kWe installed.[4,8,9,10] The solar tower concept with thermal storage has been estimated to cost between $5,140-$9,090 per kWe installed. These results have a large range which is dependent on the assumptions of the design choice. The data does seem to suggest that there is more uncertainty with the solar tower concept since there is less data upon which to base conclusions.

The solar-Stirling concept is in an earlier stage of development than the other thermal options, so the costs are rather high. However, a demonstration-sized modular plant built at the 3 MWe size (without any storage capability) is projected to cost from $4,000-$6,800 per kWe with a cost of electricity from 19.7-44.3 ¢/kWh.[4]

Solar Energy

Solar PV costs have dropped in the last several years. Capital cost estimates range from $4,750-$6,050 per kWe installed.[10] But very recent projects have been getting close to $3,000 per kWe installed.[11]

Residential Solar Economics

In 2003, the price of a solar module for residential use was about $3,200 per kWe. However, the actual installation cost on the roof, including electrical conversion and wiring, ran between $6,000-$8,000 per peak kWe.[2] Peak kWe refers to the maximum power the cell can produce under optimal sun conditions. Even in very sunny areas, the average availability of sunlight is less than 25%, which means that the power produced from the PV array averaged over a day is equal to one quarter of the peak rating.

Assuming a best-case installation cost of $6,000 per peak kWe, 25% availability, and local electricity costs at 9 ¢/kWh, then it would take about 30 years to pay back the cost of installation. The electricity cost of 9 ¢/kWh is what I currently pay in New Mexico, but it could vary in different parts of the country. Also note that in northern areas of the country that do not receive as much sun, the availability may drop to 10 or 15%.

Solar water heating is another example of residential use of solar power. A solar water heater usually costs $2,500-$4,500 to install in a typical home. The average family may use somewhere around 5,300 kWh of electricity just for water heating.[5] If I assume that solar water heating could save 75% of the energy required for water heating (3,975 kWh), and if I use a typical price for natural gas heating of $0.0463 per kWh, then a solar water heater would save the average family $184.23 per year. Assuming an installation cost for an average family home at $3,500, it would take about 19 years to pay back the cost of installation.

The long payback time still makes rooftop PV or solar water heating undesirable for the majority of the population. Considering that the average homeowner stays in his or her home for less than 10 years, most people cannot justify the expense. Realistically, for this reason, only a very small fraction of the population, who are willing to pay the high upfront cost to reduce energy consumption, will practice rooftop PV and solar water heating.

Acceptability

Solar power is probably the most acceptable form of electricity production. It does not produce any pollution while in operation and life cycle emissions are small. Certain solar PV technologies use toxic chemicals in the construction of the arrays, but it is relatively easy to prevent release into the environment. Solar plants are also fairly pleasing to the eye. Everyone I know would support solar power if it were an economical energy option. Maybe in the future there will be an environmental group protesting a solar tower since the focus area on the tower may pose a problem if birds fly into it—who knows? But for the most part, solar power will always have a great deal of acceptance amongst the general population.

Conclusion

Although solar power is one of the most accepted and cleanest forms of electricity, the problems with reliability lead to high generating costs. The solar trough concept is the most economical of the solar options if a new plant were built right now, yet even this option is expensive for large-scale use. Solar PV appears to be gaining ground by reducing production and installation costs. Only state renewable energy portfolio standards and strong tax subsidies have allowed for the construction of new plants.

Solar may have the best chance of getting into the market by producing peaking power in parts of the Southwest that see the highest number of sunny days and that rely heavily on summer air-conditioning. If electricity prices rise in the summer months, solar power may become more competitive, but it will still take significant government support to get the first plants built. The new solar trough and PV plants being planned are encouraging and may be the first commercial step to achieving reduced costs. All in all, solar energy is unlikely to make a significant impact on the entire country's energy needs in the near term, but it can play a small role.

Chapter 6: Geothermal Energy

GEOTHERMAL ENERGY involves using the heat of the earth to produce power for heating or electricity. It is classified as a renewable energy source, although this is not strictly true since energy is removed from the earth and not returned. However, since the energy content of the earth is so large, for all intents and purposes the use can be considered renewable. The United States currently has a geothermal electrical generation capacity of about 3,100 MWe.[1] Current sources of geothermal power rely on hydrothermal sites where hot water exists at shallow depths. But geothermal energy exists everywhere in the country at deeper depths and can be extracted by drilling and pumping water down to extract the heat.

Currently, a total of about 7,000 MWe of new generating capacity is being considered in the United States, and much of that work has been partially funded through the American Recovery and Reinvestment Act.[1] The resource potential is enormous, and if that potential can be tapped economically, it could provide a good source of base load generating capability. The economic challenges are considerable though.

Geothermal Heat Pumps

Geothermal heat pumps are a residential application of geothermal energy that can reduce energy use for home heating and air-conditioning. These heat pumps take advantage of the fact that the temperature of the earth just 5-10 feet below the ground remains fairly constant at about 50-60 °F year round. In the winter the earth is warmer than the outside air, while in the summer the earth is cooler than the outside air.

A geothermal heat pump consists of the heat pump unit in the home and a heat exchanger installed in the ground. The heat exchanger is simply a long pipe loop through which water can recirculate to absorb or eliminate heat. In the winter, the water absorbs energy from the earth to help heat a home. In the summer, heat from the home goes into the earth to reduce the amount of energy required for air-conditioning. Geothermal heat pumps are more suited to areas that have temperature extremes in the winter and summer.

A geothermal heat pump costs more than a traditional gas furnace and central air conditioning, but this technology can reduce energy use for heating and cooling by 25-50% each year. Reduced heating and cooling bills can make up for the initial cost in 5-10 years. People install approximately 40,000 geothermal heat pumps in the United States each year. [3]

Every location in the country is different, so when considering installing a geothermal heat pump, it will be important to work with a local heating and air conditioning company to determine what the cost savings will be. Installation costs will vary, and different areas may have different types of rebates or tax breaks. However, it can be an excellent option for reducing energy use when it comes time to replace an aging furnace or air conditioning system.

Geothermal Technology

There are two possible methods for tapping into geothermal energy on the utility scale. In rare cases, a source of hot underground water or steam can be used directly as an energy source. This is referred to as a hydrothermal source and includes geysers, hot springs, or other sources that do not reach the surface. Most existing geothermal sources of electricity are hydrothermal. If water is not present but a hot near-surface condition exists, water could be pumped down to extract the energy. These technologies are referred to as

Enhanced Geothermal Systems (EGS), and they could have potential in the future if costs can come down more.

Hydrothermal Sources

A number of naturally occurring hydrothermal sources exist around the country. Some exist in areas that are protected and will not be developed (we would not want to turn Yellowstone National Park into a geothermal power plant). Since energy efficiency increases with temperature, sources of very hot pressurized water (up to 300 °C) are the most desirable. Often this is not the case in natural geothermal sources, but as long as the temperature of the water is greater than 90 °C, the source can be used to drive a heat cycle.

Steam cycles are the most economical as long as the temperature of the geothermal resource is greater than 175 °C.[2] Direct steam from a geothermal source can be used to drive a turbine to produce electricity, but this is very rare. More common, the hot water from the resource is flashed to vapor in a lower pressure tank to drive a turbine. Figure 8 shows an example of a flash steam cycle.

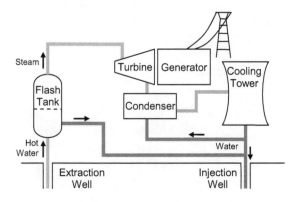

Figure 8: Flash Steam Cycle

Geothermal Energy

The flash tank takes advantage of the fact that the boiling point of water decreases as pressure decreases. When hot water from the extraction well enters the flash tank, part of it will turn into steam that can be used to drive a steam turbine. The leftover water will be returned to an injection well to replenish the water in the geothermal source.

At lower resource temperatures, the hot water can transfer its heat to a low boiling point fluid to drive a turbine. This is referred to as a binary cycle since two fluids are used (see Figure 9). The working fluid is usually a hydrocarbon. In this cycle, the hot water simply transfers its heat to the hydrocarbon to vaporize it—the vaporized coolant then drives a turbine to produce power.

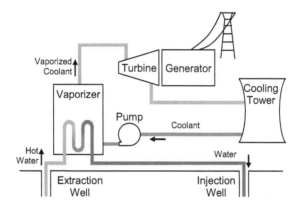

Figure 9: Binary Cycle

A concern with tapping natural geothermal sources is that underground reservoirs are not limitless. Eventually, the resource will be used up, but care can be taken to extend the life of the source. Reinjection of water into a reservoir can extend the life. In fact, some of the existing U.S. geothermal generation has dropped about 600 MWe due to declining resources.[2]

Enhanced Geothermal Systems

Enhanced Geothermal Systems (EGS) have been investigated as a way to both expand existing geothermal sources and open the door to new sources. The goal with EGS is to create a hydrothermal source in an underground location that is hot, tectonically stressed, and fractured.[4] Pressurized water can be pumped into these areas to remove the heat. The injection of the water also serves to initiate further fracturing to extend the size of the underground reservoir. (This process is commonly used in the gas and oil industries.) Figure 10 shows how the process works. Past work on EGS has shown that it is possible to create a hot underground reservoir in just about any location at any depth desired.

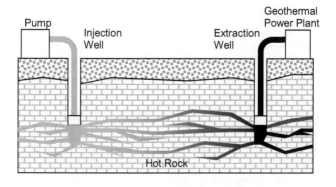

Figure 10: EGS Technology[5]

EGS depends on the formation of fractures in the ground to increase flow and surface area. The pressurized injection of water forms these fractures, which join other naturally occurring fractures in the ground. More fractures lead to more volume for the reservoir and more heating surface area, which helps to drive down costs.

A couple of hours north of San Francisco is the world's largest geothermal electrical plant called the Geysers. In 1987

Geothermal Energy

it reached peak production with the ability to provide power to 1.8 million people. However, due to declining water over the years, the Geysers currently can provide power to only about 1.1 million people.[5] To help retain this power source, waste water from the city of Santa Rosa is going to be pumped into the ground to help replenish the source. This is an example of how EGS can be used to extend existing sources.

To create new sources, EGS can be used by drilling deep into the ground to reach ideal areas that contain geothermal energy. Drilling for EGS typically requires going at least 3 km to reach 150 °C temperatures, though deeper drilling will result in higher temperature yields (and higher electrical conversion efficiency). Between the depths of 3.5 and 7.5 km the geothermal energy in the continental United States amounts to about 130,000 times our current total yearly energy demand, a huge potential of energy to tap into. The most optimal locations in the United States for developing EGS are in the southern Rocky Mountains including parts of Oregon, Idaho, California, Nevada, Utah, Colorado, and New Mexico. These areas have good potential for EGS heat around drilling depths of 5 km or so.[4]

The key difficulty with using EGS for new sources of geothermal energy centers on the high cost of drilling. It is also difficult to maintain a recirculation rate high enough to be economically useful. For very deep drilling, the power required to pump the water back up from the reservoir can be a significant portion of the power generated. It is desirable to use locations for which the heat production is strong enough to partially drive the hot water up with natural buoyancy. In past experience it has been difficult to reach high enough circulation rates to be viable.[4] Future work will need to bring down drilling costs and investigate new ways to increase circulation.

Reliability

A key advantage of geothermal power is that it provides a continuous source of renewable energy—power can be extracted 24 hours a day, and it is not affected by weather conditions. This allows geothermal plants to produce continuous base load power as opposed to solar and wind, which are intermittent. Geothermal plants are also extremely reliable with plant capacity factors often greater than 95%.[2] For this reason, it will be important to continue working to drive down costs of EGS technology.

Environmental Impact

Although geothermal energy does not burn a fuel like fossil fuels, there can be some emission release with hydrothermal sources. The most common are CO_2 and other gases that come from the underground water source. Binary plants have less of a problem with this because the hot fluid is reinjected into the ground. For the most part, these emissions are very minor and not a concern.

The biggest concern for developing natural hydrothermal sources is that they often occur at landmarks that will not be able to be developed. However, there are far too few of these sites available to provide much power anyway.

Deep drilling can cause water pollution if dissolved minerals poison near-surface fresh water aquifers. These problems are easily prevented with proper planning. Another problem with geothermal power is that it produces a great deal of thermal pollution. Because the working temperature is relatively low (for a power plant) the overall efficiency of a geothermal system may range from 10-20%. This means that 80-90% of the energy removed is rejected to the atmosphere. In most cases, cooling towers will be required to reject this heat into the atmosphere as opposed to a body of water.

Geothermal Energy

One unique concern with EGS is induced seismicity.[4] For very deep drilling and injection of water under pressure, it may be possible to induce seismic activities that could cause minor earthquakes. It is important to adequately understand the geology of a site to prevent these types of problems. Also, possible sites for EGS must have a good water source nearby. Water must be pumped down into the reservoir, and water is used for cooling. EGS will not make sense in areas of the country with limited water resources.

Economics

The development of geothermal energy can be financially risky since it is often difficult to get an exact value of the heat capacity of a potential site. With EGS technology, the risk is even higher since the technology is relatively immature. The geothermal resource can be somewhat different than other sources of energy in that the resource can be used up over time. A typical economic assessment may assume that the resource lasts for 30 years, and periodic redrilling or restimulation of the resource will be required every 7 to 10 years.[4]

Hydrothermal Sources

Hydrothermal costs are dominated by the initial development. Once operational, the fuel and operation costs are minimal. Steam plants greater than 5 MWe tend to be more economical at $1,400-$1,500 per kWe for a good site, while binary plants tend to cost more at about $2,000-$2,100 per kWe.[6,7] The development for a good resource may have a levelized cost of electricity around 5-8 ¢/kWh.[2,6] In poor sites or for smaller resources the cost will be much higher.

The cost of developing a hydrothermal source is dominated by drilling, which may account for 30-60% of the project, depending on the site. The construction of the plant may range from 30-50% of the project cost. Operation and maintenance as well as surface exploration contribute only a small amount to an overall project.[8] Although more hydrothermal sites are being considered for development, there is a limit to naturally occurring sites. The only way large-scale expansion of geothermal energy can occur is through EGS technologies that will be significantly more expensive.

Enhanced Geothermal Systems

Initially, EGS are expected to cost much more than existing hydrothermal sources. Overnight capital costs are expected at about $5,000 per kWe with the potential to drop to about $2,700 per kWe by 2030.[7] A recent study on EGS potential calculated an initial levelized cost of electricity from 10 to 30 ¢/kWh for a number of potential sites, although some sites were considerably more expensive.[4] It is evident that EGS technology is highly dependent on the location and resource available.

There are a number of existing gas and oil wells in areas of Texas, Louisiana, Alabama, Mississippi, and Arkansas that reach down to temperatures from 100-200 °C.[9] These locations may be ideal sites for use for geothermal energy. Many of these wells already produce large amounts of hot water in flow rates large enough for power production. Since the well and pumping infrastructure are already in place, the only upfront costs will be for the building of the actual power plant. Due to increasing demands for renewable energy, it is likely that these sources will be tapped in the near future. It is ironic in a way to turn gas and oil wells into a source of clean, geothermal energy in a way that keeps costs reasonable.

Geothermal Energy

Acceptability

Geothermal energy appears to be a highly accepted power source, but the costs hold back development. Geothermal energy has a very small environmental impact, but care will need to be taken to ensure that future drilling will not cause any problems with local water tables or with induced seismic activity. The ability to produce emission-free power with a high reliability makes it a renewable energy source with a great deal of potential.

Conclusion

Growth in geothermal power may double in the next few years due in large part to government subsidies. The use of hot water from existing hydrocarbon wells with EGS technologies can grow the use of geothermal power as well. Still, these technologies will probably contribute only a small amount of the needed electrical generation in the near term. EGS for new sites is currently very pricey, but the technology has a lot of room to mature.

One advantage of EGS is that it does have a good development path. The use of EGS technologies to extend existing geothermal sources will give the industry and researchers valuable experience in learning cost-saving techniques. More research and demonstrations should be encouraged to better determine the economic potential of this resource. Since drilling is the largest part of the cost, new methods for decreasing cost will need to be examined.

A key advantage of geothermal power is the reliability issue—it produces electricity all day long as opposed to other renewables like wind and solar. This means it has the greatest potential for use in generating base load power.

Chapter 7: Hydroelectric & Ocean Energy

POWER FROM RIVERS AND OCEANS represents an enormous potential across the world. Hydroelectric energy is our largest source of renewable electricity, and as a mature technology it can produce power cheaply. Ocean power technologies are relatively immature and will require a significant amount of research and development for widespread use. All of these power options are unlikely to play a major role in energy growth in the near term for different reasons. The challenges with hydroelectric in our country are a limitation of potential sites and environmental concerns. The challenges with ocean power are that these options are still too early in development and expensive to be competitive. However, the ocean could be an important energy source once the technology matures.

Hydroelectric Energy

Hydroelectric energy is an important part of our energy mix today, and it could grow the most in developing countries in the future. In North America, the installed hydropower capacity is 160,000 MWe, but it has been estimated that about 55% of the available sites have already been developed.[1] Therefore, hydroelectric is not an energy source we can depend on for much continued growth. Added to that, environmental concerns about the effects of dams and

Hydroelectric & Ocean Energy

reservoirs on local ecology and local populations have made new construction challenging. In fact many dams are currently being demolished for these reasons.

Hydroelectric plants require the construction of dams for two purposes. The first is to raise the water level to increase the potential energy available. The second purpose is to provide a reservoir to ensure that the water flow through the hydroelectric plant stays constant even in times of drought. Hydraulic turbines convert the energy of moving water into rotational energy to drive a generator.

Environmental Impact

The building of dams can disrupt local ecology in a number of ways. Flooding removes natural habitat, and the dam itself disrupts fish migration. These are two of the main concerns that have limited expansion in the past. Also, many rivers are in areas that will be protected as natural landmarks.

Although it may not be intuitive, hydroelectric energy can have significant lifetime greenhouse gas emissions. The emissions are mostly due to the construction of the dam. Poor planning can lead to other releases. The decay of vegetation under water (right after the water starts backing up) releases methane, which is much more harmful as a greenhouse gas than CO_2. Without proper planning, a hydroelectric plant could actually release more lifetime greenhouse gas emissions that an equivalently sized fossil plant. Steps can be taken to minimize this release such as clearing most of the vegetation before the dam is completed.

Economics

Hydroelectricity can be one of the cheapest forms of energy. The difficulty with hydropower is that it has a very high upfront cost, which makes it challenging to build at new sites. Investors typically want a 10- or 20-year return on investment, so in those timeframes it can be difficult to recoup the initial investment cost. In reality, hydroelectric sites can have a lifetime of 50 years, and with refurbishment, that can easily extend out to 100 years. Older dams produce power very cheaply after the initial investment has been paid off.

Capital costs may range from $1,500-$3,000 per kWe, depending on the location and size of the plant. For new installations (in the first 20 years) the cost of hydroelectric power may range from 6 to 10 ¢/kWh.[2] Once the loan has been repaid, the costs drop considerably. In some cases, hydropower from older plants costs below 1 ¢/kWh, the cheapest cost of any available option.[1]

Ocean Energy

In the future the oceans very likely could be the next step in human exploration and expansion. Our use of the ocean is mostly limited to coastal areas now, but one day the oceans could be a critical resource for food, mining, and energy. Research has only begun to break the surface (pardon the pun) of ways to extract energy from the ocean.

There are three different types of ocean power technologies that are promising as energy sources. Currently these technologies are expensive, but some may be able to become more competitive as the technologies mature. Tidal power uses the difference in the water height from high to low tide to generate energy. Wave energy conversion extracts energy from the oscillating motion of waves. Lastly, the energy from ocean currents can be extracted with underwater turbines.

The advantage of ocean power sources is they are much more constant and predictable than other types of renewable energy. As such, it would be easier to incorporate these power options into the grid. Ocean power will obviously be useful only to coastal states, but since many of our largest cities are on the coasts, the generators could actually be located close to the high-demand areas. On the other hand, ocean energy systems must be robust enough to survive the worst storms, which can add to the cost. Construction of systems on the seabed is expensive and offers different challenges.[3] These are the two major technical challenges that will be overcome only with more commercial development.

Tidal Power

Tidal power uses the difference in height from low tide to high tide to generate electricity from the ocean. In effect, tidal power taps into the gravitational energy between the earth and the moon. In a simple type of tidal power station, a barrage (dam) is built across a river estuary to trap water at high tide. At low tide the water is released through turbines to produce electricity. A number of sites have been examined over the years for tidal power, but this option is usually too expensive to be competitive. Only a handful of tidal power stations have been built around the world.

Due to the change in water flow patterns, tidal power will have environmental consequences mostly related to fish migration and sediment flow. Similar to hydropower, environmental concerns will make these projects challenging in the future if the costs are able to come down. There also are few sites suited to this development.

The reason for the high cost of tidal power is that construction in the marine environment is difficult, and the difference in water height is small. The building of the barrage dominates the cost of electricity. One advantage is that the lifetime of the barrage can easily be over 100 years. There is limited data available on costs, but the costs have been found to be around $1,800-$2,000 per kWe.[4] A further difficulty is the low capacity factor around 22-35% since the station can operate for only part of the day. Cost predictions have placed the levelized cost of electricity anywhere from 10-40 ¢/kWh.[1] Due to the high costs and a limitation on suitable sites, tidal power is unlikely to play a major role in our energy needs of the future.

Wave Energy

Wave energy converters are one of the more exciting new technologies for renewable energy. A number of different options are being explored for generating power from this resource. These devices may be onshore or offshore and usually involve unique engineering to convert the up-and-down motion of waves into electrical energy. Wave energy converters are expected to have little environmental impact. Since their effect is to calm the ocean, they actually help prevent coastal erosion. Some onshore systems will have an impact during construction, but that will be short-lived. The largest concern is probably just the visual impact that it may have on the water.

Wave energy offers higher energy density than wind energy, which means more energy can be extracted with smaller devices. However, the state of development of wave energy is at a similar point to where wind was 20 years ago.[5] If wave energy technology can be optimized like wind turbines have, it could provide a valuable energy source in a couple of decades. Although there are many areas along both coasts that have good wave resources, the best region for developing wave energy in the United States is off the Oregon coast where wave heights off shore can measure 3.5 m during the winter.[5]

Various techniques can be used to extract energy from waves. Onshore generators use the breaking waves to generate energy. Some techniques that are being explored include an oscillating water column, tapchan, and pendular devices.[6] In an oscillating water column, wave action causes a column of water to rise and fall, which in turn pressurizes and then depressurizes an air column. The pressurized air is used to drive a turbine. In a tapchan (short for "tapered channel") water is fed into a reservoir above sea level—the stored water then travels through a turbine to produce energy. Finally, a pendular device uses the wave action to rock some type of flap back and forth—this motion powers a generator.

Offshore systems use the up-down motion of waves to produce power. The direct drive technique involves attaching a buoy to some type of generator—the buoy may either be floating or completely submerged. Permanent magnetic linear generators can transform the up and down motion of the buoy into electrical power. Or the buoy can be used to pump water through a turbine. Other techniques use the rise and fall of waves to alternately stretch and relax a hose to push water through a turbine. Figure 11 shows some examples of different wave-energy converters.

Figure 11: Wave Generator Concepts: (A) "Pelamis" Prototype (Courtesy of Pelamis Wave Power Ltd.)[7], (B) Permanent Magnet Linear Generator Buoy (Courtesy of Oregon State University, illustrated by Nicolle Rager Fuller, NSF)[8]

Environmental challenges are usually minimal and include only the impact of the initial construction. The biggest challenge will be in finding sites that are not on scenic shorelines.

Wave energy converters are still pricey, which is why there are a number of different options being investigated. However, these costs may have a decent chance of coming down and finding small-scale uses. Costs for a commercial-scale modular wave-energy plant capable of generating an average of about 35 MWe have been estimated at 9.2-11.2 ¢/kWh for a good coastal location in the United States.[9] Capital costs appear to be dropping as new projects are built around the world. Current costs are actually less than the average cost of wind power when the installed capacity of wind turbines was about equal to the installed capacity of wave energy converters today. In other words, as the installed capacity of wave energy increases, costs are expected to decrease substantially.

For the near term, it is unlikely that wave energy will make a major impact on our energy demands. However, wave energy converters could be very competitive in about 20 years if the technology is pushed. This energy option is one of the most nonintrusive power options with very little environmental impact. It is also a more predictable form of energy than wind and solar. For all of these reasons, it would be advantageous for the United States to support a strong research and development effort in this area.

Ocean Current Energy

Ocean currents are another form of energy that can be tapped using turbine technology. Any strong flow of water can be used to generate energy with a turbine. The turbines look similar to a wind turbine and may be either mounted to the seabed or mounted to a floating platform for deeper waters.

Ocean current energy is also at an early stage of development, and grid-connected turbines are only beginning to be built. Further research will be required to get better cost estimates.

Hydroelectric & Ocean Energy

Figure 12 shows an example of a prototype in operation. There will be challenges unique to underwater operation including prevention of cavitation (bubble formation), prevention of marine growth, corrosion resistance, and increased reliability.[10] Other economic challenges include the cost of cabling to transport the electricity to the shore.

Figure 12: Prototype Design for an Ocean Current Turbine (Courtesy of Marine Current Turbines TM Ltd.)[11]

Ocean currents are very predictable and more constant as compared to wind energy. This can make it a more reliable form of renewable energy. And similar to wave energy, the energy density in ocean currents is much higher than in air. Potentially this means that for a similar-size turbine, ocean currents will produce much more power than wind. Finally, since ocean current energy systems are all submerged on the seabed, there is no visual impact of this technology (which could be a major advantage over wind turbines). Ecological impact will be an issue, but turbines can be designed to move slowly to allow fish to pass, and cages can be used to prevent damage to larger species.

Similar to wave energy converters, it will take many more years of development to bring down costs and get a better idea of the technology's potential. Again, this is an energy source that will not make a major impact in the next 20 years, but it could be an important renewable energy source soon thereafter.

Conclusion

Due to a lack of potential new sites, hydroelectric and tidal power stations are unlikely to grow much in the future. Their environmental impact is a concern in a time of trying our best not to disturb natural ecology. Wave and ocean current energy may have the most potential for future use, but for the time being these technologies need to mature more to make a major impact on our energy demands. However, due to their long-term potential, it would be advantageous to support ocean energy technologies with a strong research and development program.

Chapter 8: Biomass Energy

IN LOOKING BACK on the progression of energy sources throughout history, for a long time wood was the primary energy source. Biomass is simply a green-sounding name for a return to extracting energy from wood or crops. This is not the first time that a simple name change has been used to make an industry sound better than it is. It is similar to how the term "garbage collection" is not used anymore—instead "waste management" is the preferred term.

Biomass encompasses a number of different ways to extract energy from plant matter whether by direct burning or by converting a crop into a different fuel. Corn ethanol falls under this heading. Standing forests may be used, but natural forests grow too slowly to be of much use as a sustainable energy source. Biomass may also come from organic wastes, which include municipal waste, industrial wastes like pulp or wood chips, or nonfood agricultural wastes like straw or husks.

Other than hydroelectric power, biomass is the largest source of renewable energy in the United States, contributing about 4% to total energy consumption.[1] The majority is from wood, with smaller contributions from waste or ethanol. I use the term "renewable" because the feedstock can be regrown for a sustainable energy option.

Many environmentalists support the use of biomass, which never made a lot of sense to me. The justification is that the growth of the crops will absorb just as much CO_2 as is released in the burning process. Therefore, burning biomass releases zero net carbon emissions. (Fossil fuels, on the other hand, release CO_2 which was locked away in the earth as hydrocarbons.) But biomass does not reduce emissions either, and the burning still produces pollution in the local area. We should focus more on technologies that can make a dramatic difference in carbon emissions.

There are some small-scale uses where it does make sense to extract energy from what would otherwise be a waste product. But the intentional growing of biomass takes up valuable farmland and can lead to other environmental issues like fertilizer and pesticide pollution. The land requirements to produce significant quantities of biomass energy will be enormous and unpractical.

Biomass Technology

There are two different methods available for using biomass. The fuel can be directly burned to produce heat to drive a power plant, or the feed can be converted into another fuel using some type of thermal or biochemical method. This can include gasification or liquefaction. Chapter 2 already examined many of the issues associated with the use of biomass to produce biofuels, so the focus of this chapter will be on the direct burning of biomass.

Energy crops will be required for a more widespread biomass industry. Essentially, this just makes the use of biomass more efficient by growing the fuel to be burned or converted over and over again. The goal with direct burning is to use a crop with the highest energy content that can be re-grown quickly.

A strong limitation on the use of biomass will likely be the amount of available land. In some cases, as in the United States, an overproduction of certain crops like corn could instead be used as an energy source. On the other hand, it

Biomass Energy

does take up valuable farmland that may be better used to produce food for an expanding population. It will take roughly 1.8 square miles of land to provide enough energy crops to support a direct-fired 1 MWe biomass plant.[2] If it were feasible to build, a large 1,000 MWe biomass power plant would then require 1,800 square miles of land! For this reason, biomass is not economical as a large power plant.

Direct Burning

Biomass can be directly burned in a boiler to generate steam and drive a turbine. Biomass burners tend to be smaller in size in the tens of MW range, which limits the conversion efficiency to 20-25%. However, improvements in drying the biomass and increasing the size of the plant could increase efficiencies to above 30% if the application is developed more.[3]

Biomass can also be co-fired in a coal power plant to displace some coal while taking advantage of the higher efficiency of a large plant. Up to 2% biomass could be substituted for coal without any modifications, but up to 15% could be substituted with minor plant alterations.[3] However, these small changes in the amount of coal burned will not make much of a difference in overall carbon emissions.

It is possible to gasify biomass so that it will burn more efficiently. The biomass feedstock is partially combusted in air or oxygen to produce a gas of carbon monoxide, hydrogen, and other combustible compounds. This gas can then be burned with efficiencies as high as 45%.[3] However, this technology is at an immature point in development.

Biomass Digesters

Biomass digesters are much different from direct burning and work by converting the material into methane. When organic material decays in the absence of oxygen, methane is produced that can be collected and used to generate electricity. Landfills can produce large amounts of methane in this fashion, which can be used to drive a small power plant, some as high as 20-30 MWe.[3] This technology can also be specially designed to produce methane from animal wastes or human sewage.

The production of natural gas from biomass is an excellent way to produce energy from what would otherwise be a waste. It also can be an economical source of energy. I have always been a strong supporter of recycling whenever possible, although in many cities the infrastructure for recycling is not well developed. An extensive recycling program for glass, metal, and paper products along with biomass digesters for organic wastes would be a much more efficient use of resources while at the same time drastically reducing waste. Although somewhat outside of the scope of this book, I hope these types of recycling technologies will be more strongly supported in the future.

Net Energy Analysis

The net energy analysis for the direct burning of biomass will be better than that for corn ethanol. Direct burning of woody crops does not require an energy input to convert the fuel. However, it does still require energy to grow, harvest, and transport the crops to the biomass plant. The fewer fertilizers that can be used for the production of woody crops, the better. The net energy ratio for the growing of poplar, sorghum, or switchgrass was found to range from 10-15 for direct burning applications. In other words, the burning of the fuel produces about 10-15 times more energy than the energy inputs required for farming, harvesting, and hauling.[4]

Environmental Impact

The immediate CO_2 emissions from the direct burning of biomass have been estimated at 939 g/kWh for an integrated gasification combined cycle or as high as 1,217 g/kWh for a traditional boiler.[5] These values are higher than coal, but if it can be assumed that the growth of the crops reabsorbs as much CO_2 as was released during the burning, the life cycle emissions are very close to zero. The direct emissions are an important consideration since they do present a pollution source to the immediate area, potentially making a large biomass plant undesirable to local populations.

Direct burning of biomass releases other compounds into the atmosphere like the burning of fossil fuels, including particulate material, carbon monoxide, and nitrogen oxides. These gases may need to be controlled depending on the system. The one difference from coal burning is that biomass does not contain sulfur or heavy metals, so expensive scrubbers will not be required.

Economics

The direct burning of biomass to produce electricity has a much lower efficiency than coal, so it costs more. Most of the current economical uses of this technology occur when an industry directly uses its waste for heat or power. Dedicated biomass plants with energy crops are more expensive. More improvement will be needed on the overall efficiency using techniques like gasification to bring down costs.

Capital cost estimates for a stand-alone biomass plant are around $3,860-$7,900 per kWe.[6,7] Estimates for the levelized cost of electricity vary quite a bit from 6.2-11.6 ¢/kWh, though the higher end of this range is probably more realistic. A study examining the potential of using biomass for woody crops in Spain found a levelized cost of electricity ranging from 9.2-11.1 ¢/kWh for direct burning or 8.2-8.8 ¢/kWh for biomass gasification followed by a more efficient cycle.[8] These costs were projections for large-scale use and were probably more of a future cost since this technology requires more research and development.

Acceptability

Biomass has in the past been well accepted by most people, mostly because of the belief that it is better for the environment than other sources of energy. As more is learned about the real environmental impacts of biofuels like ethanol, these options are likely to lose support. In some instances the burning of biomass or production of biofuels is not desirable because of unpleasant smells that may be generated in the area. On the other hand, the direct burning of organic wastes may be more accepted as a way to reduce landfill use.

Conclusion

In the best-case scenario, the purpose of biomass is to keep carbon levels in the atmosphere constant. But with a growing population and a diverse energy mix, emissions are only going to increase, and biomass does not help to reduce emissions. It would be much more logical to push for truly zero emission plants while using available land to restore forests or grasslands to absorb carbon without reemitting it. Due to the large land requirements for biomass, it just would not be possible for biomass to be used as a major energy source in the future.

Biomass use for direct burning may see only limited growth in the near term due to the economics. The cost of building much larger-scale plantations to fuel power plants drives up the fuel cost. But biomass will continue to have a small role in power production, and can be useful in burning or converting human or industrial organic wastes to reduce landfill requirements. It makes much more sense to use biomass in this fashion as a way to conserve resources by generating energy from what would otherwise be a waste.

Chapter 9: Wind Energy

WIND ENERGY has made a great deal of progress in the past 20 years and is likely to be the renewable energy option that will expand the most in the next couple of decades. More than 15,000 wind turbines in California and 2,800 in Denmark have given the industry extensive experience in design, operation, and maintenance.[1] With modern materials and increasingly larger turbines, the cost of wind energy has become economical in areas with good wind resources. Continued improvements must be made to build turbines that are economically competitive in areas of lower-quality wind resources. The continued expansion of wind energy will allow for more of these cost improvements to be achieved.

The difficulty with wind energy is mostly due to a low reliability. The intermittent availability of wind makes it difficult to depend on as a base load source of power. Siting, though, has proven to be a barrier in some areas of the country, mostly due to public concern over visual impact. Later in the chapter I describe an example of public opposition to a wind farm.

When I was seven years old, I remember driving through the wind farms in California on a family vacation. It was a beautiful day for a drive, and the wind turbines were such an amazing thing to see. It disturbs me that some areas of the country would resist building them due to visual impact. I wish all people could appreciate the sight of a wind turbine for producing clean energy as opposed to viewing it as an eyesore.

Despite these concerns, wind energy is one of the fastest growing renewable energy technologies. From 2000 to 2009 electricity generation from wind grew by a factor of 12 bringing the total to about 35,000 MWe.[2] It will likely continue to grow due to state mandates requiring more development of renewable energy and because wind is such an environmentally friendly option. Areas with lower wind speeds are about 20 times more prevalent than high-wind-speed areas and much closer to cities that demand the power,[3] so if costs continue to decrease, wind could contribute a sizeable fraction of our power generation in the future.

Wind Turbine Technology

Wind turbines have gradually become larger over the past few decades because of the increased efficiency that larger sizes offer. Designs in the 1970s were in the 30-60 kWe range, with sizes progressing to between 300 and 500 kWe in the 1980s. Currently 5-10 MWe turbines are being designed.[4]

The design of a wind turbine is simple. Rotors are used to transform wind energy into rotational energy. A gearbox and generator are then used to convert the rotational energy into electricity. The engineering of wind turbines is all about optimizing the design using lightweight materials for the rotors and reliable parts for the gear box and generator.

The standard wind turbine design has a rotor attached to a horizontal shaft with a yawing system to rotate the blades into the wind (see Figure 13). The rotor blades can rotate at a near-fixed speed or at a variable speed depending on the design. The rotor is raised on a tower to be clear of the turbulent layer of wind near the ground. The gearbox and generator are attached directly to the turbine shaft at the top of the tower. This gearbox must be extremely rugged to withstand fluctuations in wind speeds, turbulence, and bending forces on the rotor blades.

There are also vertical axis wind turbines that look quite a bit different and can operate with the wind blowing from any direction. Several designs have been tested, but they have not achieved the commercial success of the traditional horizontal design.

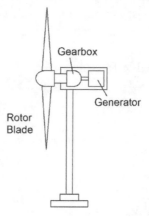

Figure 13: Horizontal Axis Wind Turbine

The rotor blades are typically made of a lightweight material like glass-reinforced plastic, wood epoxy, or carbon fiber. Two- and three-blade designs are used for cost effectiveness and smooth operation. The rotor size is set by the desired output of the turbine—a 600 kWe turbine will have a rotor diameter from 40-50 m, while a 5 MWe turbine may have a rotor diameter of 120 m.[4]

Siting Wind Farms

Wind turbines are usually deployed in groups as wind farms. The turbines are spaced far enough apart to prevent interference from one to the other. The power from the farm must be delivered to the local power grid, which requires a substation. Larger wind farms require frequency and voltage stabilization, which can add slightly to the cost.

The potential of the power of a wind resource is divided into seven classes as shown in Table 3. The wind speed (in meters per second) and wind power density (in watts per square meter) are given for both 10 and 50 meters above the ground. Turbines in Class 5 and above are currently economical in most areas of the country, but it is hoped that turbines will be economically competitive in Class 3 and above in the next few years. The power output from a wind turbine increases substantially with wind speed, so small changes in speed lead to large changes in power output.

Wind Power Class	Wind Speed at 10 m (m/s)	Wind Power Density at 10 m (W/m^2)	Wind Speed at 50 m (m/s)	Wind Power Density at 50 m (W/m^2)
1	0-4.4	0-100	0-5.6	0-200
2	4.4-5.1	100-150	5.6-6.4	200-300
3	5.1-5.6	150-200	6.4-7.0	300-400
4	5.6-6.0	200-250	7.0-7.5	400-500
5	6.0-6.4	250-300	7.5-8.0	500-600
6	6.4-7.0	300-400	8.0-8.8	600-800
7	7.0-9.4	400-1000	8.8-11.9	800-2000

Table 3: Wind Classes[1]

The wind energy potential in the United States is enormous. Research finds that even with strong limitations on the land that could be used, the land areas with Class 5 and greater resources could provide an additional 60-80 GWe of new generating capacity, which could provide about 13-18% of current total electrical demand. Land areas with Class 3 and greater could produce an additional 500-1,200 GWe of new generating capacity.[1] About 80% of this potential could come from North Dakota, Wyoming, and Montana. The problem, though, is that these areas are not close to the major urban areas in the country, so extensive transmission systems would be required.

There are also many offshore locations with excellent wind conditions. The U.S. offshore resource in shallow waters could potentially provide up to 900 GWe of new generating capability.[3] About 800 MWe of offshore capability is currently installed around the world, with some plans for the United States. Wind farms are being considered for offshore placement due to the limited land areas near dense population centers. Locating wind farms at sea is advantageous because they see a more reliable and consistent source of wind with less turbulence and low visibility. Wind farms can be sited out far enough to be invisible from shore. The primary disadvantage is cost—offshore wind farms can cost from 40-100% more than onshore systems.[4] The costs for maintenance and the grid connection are also higher. Conditions at sea are more severe, which means the wind turbines need to be more robust.

The siting of wind farms ultimately will be limited by the amount of power that the local transmission systems can withstand. Grids may need to be strengthened in order to accept wind power in a remote region. For offshore sites, the coastal grid must be strong enough to accept the added power. Any modifications to the grid will increase costs.

The limited reliability of wind power also needs to be taken into account in siting wind farms, as other sources of power must be available when wind power is low. Connecting multiple wind farms spread out over hundreds of miles helps to increase the reliability, but extensive transmission systems will be required to get the power to the load centers (cities). Due to the structure of our current energy market, it can be very difficult to get new transmission lines built; so this could be a potential roadblock in developing wind energy.[5]

Reliability

Although there are a number of advantages to using wind power, reliability is still one of the biggest issues holding back its development. As the seasons change, the average wind strength changes. These yearly changes in wind speed affect costs the most (as opposed to daily or hourly fluctuations). For these reasons, large-scale use of wind energy will require backup or energy storage that is available in times of low power production. Natural gas is the easiest technology to bring up in power output quickly.[6]

As long as wind power makes up only a small fraction of the total power production, the variation in supply gets lost in the variation in demand. The highest projections show that wind power can make up at most 20% of grid energy without having detrimental effects to the electrical supply.[4,6] However, the added control features necessary to deal with significant wind power production will add to the overall cost of the system.

About 16% of the total electrical demand in Denmark comes from wind energy.[7] In fact, Denmark dominates the market in manufacturing wind turbines and has aggressive plans to continue to expand wind generation with both onshore and offshore systems. They serve as a model for the rest of the world by proving that wind can be used for a significant fraction of a country's power generation.

Additional costs on the overall system, assuming that wind generates 30% of a country's electrical generation, were found to be about 1.8 ¢/kWh for grid reinforcement, managing losses, energy balancing, and security.[5] The cost of intermittency is also dependent on the other power generation technologies in the locale. For example, natural gas and hydro have quick ramp-up rates so both can deal with intermittency better and with minimal cost. Coal and nuclear, on the other hand, have slow ramp-up rates, and cannot deal with intermittency as well.

Environmental Impact

Wind power is a clean source of renewable energy. Lifetime carbon emissions from wind power are generated only from construction of the turbines, and equal about 15 g/kWh.[8]

Blade noise can be a problem with wind turbines, but newer designs have minimized the noise to levels just slightly above the sound of the wind. Government regulations may limit the noise level and spacing to the closest dwelling to prevent it from being a nuisance. Noise concerns are of particular interest during siting studies.

Another concern of the past is the possibility of injuring or killing bird populations. These concerns are largely unjustified. Newer designs can help to mitigate this threat, and there is evidence that birds learn to avoid the turbines. Analysis of a coastal wind farm in the Netherlands found that the wind turbines were far less detrimental to birds than high voltage transmission lines and freeways.[1]

Economics of Wind Power

Wind power has steadily grown in the United States since the 1990s due to government incentives and regulations requiring utilities to provide a certain portion of renewable energy. Similar incentives across the globe have led to significant decreases in cost in the past couple of decades. At the same time, as the average size of wind turbines has increased, so too has the production efficiency improved.

Current installation costs for onshore wind farms fall around $2,400 per kWe installed.[9,10] Offshore turbines currently cost around $6,000 per kWe installed, with expectations for cost reductions in the near future.[9,10,11] The levelized cost of electricity for onshore systems is expected around 10 ¢/kWh for good wind resources.[10] The cost of electricity will be on a sliding scale with the quality of the wind resource. The levelized cost for offshore systems is expected to be around 24 ¢/kWh, again depending on the resource.[10]

Operation and maintenance costs make up a sizeable fraction of the total annual costs of wind energy, and these costs increase with time. At the beginning of life, the maintenance costs may be around 10-15% of the total levelized cost, while at the end of life, it might be around 20-30%.[6] Since wind power is relatively new, few turbines have been around for 20 years, so there is some uncertainty about these long-term costs.

Acceptability

Developers of wind turbine technology in the past have underestimated the problems of siting wind farms. The Cape Cod experience is one of the most telling examples.[12] A private company, Cape Wind Associates, is planning to build 130 wind turbines (each at 3.6 MWe) in 24 square miles of Nantucket Sound. This area off the coast of Cape Cod is an ideal site due to the strong and reliable wind resource there. Each wind turbine is proposed to sit 128 meters out of the water, but from the shore the turbine would be barely visible on the horizon. The total peak output would be 420 MWe to provide power to Massachusetts.

This proposal was soon met with a strong, well-financed opposition from a significant faction of the local population. Most of the opposition centered around concern over intruding and overdeveloping the Cape and the ocean in general, which for many in the area is almost a sacred place. There were other concerns about interrupting the view and possible damage to the ocean's ecosystems.

This opposition has placed environmentalist against environmentalist in a unique way, and it points to a downside of wind technology. Some of the best places for wind resources are large open expanses both on land and offshore, so there could be concern about developing these areas. I think that this problem points to a troubling development in our society today. Our energy demand is constantly increasing, and utilities need to build new plants. We have to accept the need for some tradeoffs and consider the alternatives. Can we live with an offshore wind farm that has a very low visual impact if it prevents the building of another coal power plant? We have got to realize that we need clean energy options, and when an industry comes in and wants to develop a clean energy option like a wind farm, we should embrace it.

Conclusion

Wind power is currently one of the most economically competitive sources of renewable energy. Due to renewable energy portfolio standards, construction of wind farms is likely to continue; they could provide a sizeable portion of our energy in the future. The technology continues to be optimized to bring down costs, but more cost reductions will be required so that areas with lower wind speeds can be used.

It is advantageous that wind farms are built in a modular fashion with a number of smaller units, so individually each turbine is relatively cheap to build and can take advantage of mass production of parts. The disadvantage is the poor reliability of wind, which requires backup sources of energy. Energy storage costs currently would make wind power too expensive, so most utilities will likely need to rely on natural gas to make up power when wind speeds are down.

Wind Energy

I believe that the siting problems of wind farms in the past have a great deal to do with taking energy for granted. If a local community decides to reject a wind farm, it will likely result in the building of a coal or natural gas plant somewhere else. These types of problems may start to go away now that climate change is taking more of a center stage in the political debate.

Part III

Nuclear Energy

NUCLEAR ENERGY has always been in a class by itself. The ability to extract the energy of the atom is so different from renewable energy or burning fossil fuels. As such, nuclear power plants have different challenges than other power options. While nuclear plants are well established and economical, public acceptability has been a key barrier to development in the United States. But nuclear plants provide about 20% of our electrical supply *without any greenhouse gas emissions*, which is why this option needs to be considered as part of the solution for achieving a clean energy future.

Chapter 10 discusses existing nuclear power plants that generate energy by the splitting of the atom. Much of the focus is on the issues that lead to uncertainty concerning nuclear energy. Chapter 11 discusses fusion energy, which generates energy from the joining of two light atoms. Although fusion research will still require many more years of development, it could be an important energy source in the future. Many of the key technical issues of fusion are introduced.

Chapter 10: Nuclear Energy

AFTER MY FIRST YEAR of graduate school, I had the opportunity to be an intern at Los Alamos National Laboratory. I worked on a small experiment that was housed in a radiation hot cell. (A hot cell is a room that is well shielded with thick walls of concrete and used to study or work with strong sources of radiation.) The experiment produced x-rays during operation, which is why I needed a hot cell, but when the experiment was turned off, it was perfectly safe to go into the cell to perform maintenance.

My family came out to visit one weekend, so I took them into the hot cell to explain the experiment I was working on. Before going in, of course, I briefed them on the necessary safety precautions. While in the hot cell, I noticed that my mom and my sister-in-law were not too eager to stay in there that long—I believe they were unsure about being in a room designed for people to work with radiation sources. Their visit gave me a better appreciation of the uncertainty surrounding nuclear technologies. I am very comfortable with radiation and nuclear-related technologies because of my background and training. But my mom and sister-in-law had never dealt with anything like that before.

I believe there is a large gap between the actual risks of nuclear power and the perceived risks, and closing this gap will be one of the challenges in order for the country to be more accepting of this power option. Nuclear power plants have excellent safety records and produce power at competitive prices. With increasing concerns about climate change, some environmental organizations are starting to accept nuclear power since it produces no greenhouse gas emissions. But there are some tradeoffs that the country will need to come to terms with for this power option to expand. As part of analyzing the nuclear power option, this chapter focuses on these tradeoffs.

Nuclear Reactor Technology

Nuclear reactors harness the energy released by the splitting or "fissioning" of the atom. When the isotope uranium-235 is bombarded with neutrons, it splits into two fission products and releases energy. The energy is released in the form of heat. The fission process also releases neutrons—these neutrons can then initiate more fission events. In a nuclear chain reaction a neutron initiates a fission reaction, then the reaction generates a neutron, then that neutron initiates another fission event, and so on. In a nuclear reactor, the reaction is controlled such that the reaction rate is held constant.

Uranium must be enriched before it can be used in reactors in our country. Natural uranium contains only about 0.7% uranium-235, and this concentration must be increased to 4-5% for fuel. The enriched uranium is fabricated into fuel rods arranged into a series of fuel assemblies (shown on the left in Figure 14). The rods are spaced apart to allow room for cooling water. A zirconium or steel cladding surrounds each fuel rod to contain all of the material during operation in the reactor.

A large nuclear reactor core may have hundreds of fuel assemblies arranged together. Once nuclear reactions are initiated in the reactor, the goal is to keep the reaction rate proceeding at a constant rate. Pressurized water flows around the fuel rods to remove the heat generated from the fission process. The reaction rate is controlled using rods that contain a neutron-absorbing material. These control rods are evenly distributed throughout the core. Insertion of the control rods into the core will absorb more neutrons, and thus decrease the power level. Removal of the control rods results in an increase in power level.

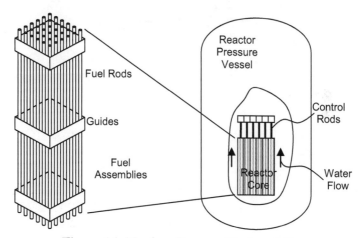

Figure 14: Nuclear Reactor Fuel and Core

During normal operation at constant power output, the control rods are adjusted only slightly to keep the reaction rate constant—much like how we move a steering wheel on a car back and forth just a little bit on a straight stretch of road to keep the car in the lane. But the control rods can be inserted fully into the core to shut down the reaction in an accident scenario or when the reactor needs to be shut down for maintenance.

Nuclear reactors in the United States are all Light Water Reactors (LWRs), but there are two types depending on the coolant design. Pressurized Water Reactors (PWRs) contain a water coolant loop through the core that is at a very high pressure to keep the water in a liquid state. The hot water transfers its heat to a steam generator to drive a steam turbine. Figure 15 shows an example of a typical PWR.

Boiling Water Reactors (BWRs) operate at a lower pressure such that the water is designed to partially boil within the reactor core. The steam from the core directly drives a steam turbine to produce power. Both PWRs and BWRs typically reach conversion efficiencies around 35%. Since 65% of the thermal energy from the plant needs to be rejected to the environment, large cooling towers are usually needed to expel the heat into the atmosphere. Cooling towers on nuclear plants simply evaporate water to remove the heat, so only water vapor leaves the top of the towers.

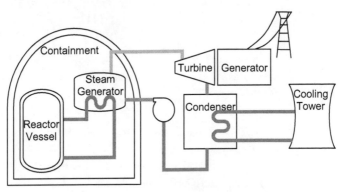

Figure 15: Pressurized Water Reactor

The thermodynamic cycle and power plant portion of a nuclear plant are really no different than any other fossil-fuel power plant. The biggest difference is the reactor vessel. The reactor vessel and steam generator are housed within a containment structure. The containment is designed to keep radioactive isotopes contained in the unlikely event of a major accident. There are also numerous safety systems, such as redundant pumps, that keep the core cooled in the event of an accident.

All current reactors utilize large designs to take advantage of economies of scale. However, a number of small modular reactors designs have been developed in recent years. Smaller designs may open up nuclear to use in small to mid-sized cities or even for alternative applications like industrial process heat. These designs may have improved safety characteristics due to smaller cores, more reliance on passive cooling, and underground siting of the reactors. Although these designs are not proven yet, they may be a technology to keep an eye on in the future.

Uranium Resources

The estimated economically recoverable uranium resources in the United States at $50 per pound are about 245,000 metric tons of uranium oxide.[1] The estimated additional resources (recoverable at higher prices) are about 2.2 million metric tons.[2] Natural uranium contains mostly uranium-238. Because the uranium-235 content must be enriched before being used in U.S. reactors, it takes about 8 metric tons of natural uranium oxide to create 1 metric ton of usable fuel.[3] Our total uranium resources would provide enough fuel to power our current fleet of reactors (100 GWe) for about 130 years.

Nuclear energy has an alternative option that extends uranium resources considerably. Alternative reactor designs can transform the abundant uranium-238 isotope into the fissionable plutonium-239 isotope. In fact, it is possible to design a reactor to create more fuel than it burns. Since there is so much more uranium-238 in the ground, these advanced reactors could provide plenty of energy for centuries, even at expanded future energy demand. Fortunately there is plenty of time to develop and demonstrate these reactors before supplies of uranium-235 run low.

Reactor Safety

The advantage of light water reactors (LWRs) is that the loss of water in the core will shut the reactor down. Neutrons produced from fission are moving at very high speeds. In order for them to be used to initiate additional fission events, they need to be slowed down or "moderated." In effect, the neutrons bounce off the hydrogen in the water coolant to lose energy. (Imagine how a billiard ball slows down as it bounces off other balls on a pool table.) If the water is not present, the neutrons cannot slow down, and the chain reaction will stop. This is an advantage because the reactor core will never be operating at full power if the water coolant is not present.

(The devastating Chernobyl disaster in Russia involved a different reactor design that used solid graphite as the moderator. The problem with this design was that in the accident, the loss of coolant did not shut down the reactor, so it continued to generate full power with no way to remove the heat. This power buildup is what led to an explosion that spread radioactive materials in the surrounding area.)

However, in light water reactors, even after the reaction has shut down, the fuel rods are still generating a considerable amount of heat from the decay of the radioactive isotopes left over from all the fission reactions. A large nuclear reactor core may generate 3,000 MWth during full power. Immediately after shutdown the decay heat may be as high as 200 MWth, and after two hours it drops to about 30 MWth.[4] This decay heat is strong enough to melt the core if the heat is not removed. For this reason it is always important to keep the core cooled in the event of any accident scenario. The entire safety system of an LWR involves backup systems and redundant pumps to keep the core submerged under water.

Two significant accidents have occurred with LWR designs: the very recent accident at the Fukushima Daiichi nuclear power plant in Japan, and the Three Mile Island accident in the United States. Both highlight the incredible importance of keeping a core cooled at all times. The following callouts briefly describe the two accidents, although details of the accident in Japan are still emerging.

Three Mile Island

The Three Mile Island accident near Harrisburg, Pennsylvania, started due to a loss of coolant from the reactor core. A relief valve got stuck in the open position and started leaking coolant water from the core. Due to poor training and poor management, reactor operators made a relatively minor incident into a serious accident by cutting off backup systems. The water level in the reactor vessel dropped enough to uncover the core. The fuel rods began to melt, releasing radioactive material into the water.[4]

Operators were eventually able to find the problem and restore cooling, but not before a significant amount of the fuel in the core melted. Some radioactive gases built up in the system and accidentally leaked into the atmosphere, but fortunately the releases led to very small doses that were not a health threat to the surrounding area. There were no injuries or detectable health impacts from the accident. The average person receives a yearly background radiation dose of about 300 millirems. The average radiation dose to people living within 10 miles of the accident was estimated at 8 millirems, so the dose increase in the surrounding area was well below any level that could cause a health concern.[7]

The Three Mile Island accident was viewed as an isolated incident, and the errors caused by the operators were not representative of the rest of the nuclear power industry. However, it served as a good lesson to prevent such an accident from occurring again. Although there were numerous errors and a small release of radioactivity, for the most part, the Three Mile Island accident proved that the numerous safety and containment systems prevented a serious release of radioactivity.

Fukushima Daiichi Nuclear Incident[5]

On March 11, 2011 a magnitude 9.0 earthquake struck and initiated a tsunami off the coast of Japan. The three operating reactors at the Fukushima Daiichi nuclear plant underwent automatic shutdown with full insertion of control rods into the reactors. However, the tsunami disabled all AC power to Units 1,2, and 3. The emergency diesel generators which are used to provide cooling for the reactors, were disabled due to the flooding.

Lack of cooling caused a hydrogen buildup in the Unit 1 reactor building, which led to an explosion on the second day. Operators started injecting seawater into Unit 1 to keep the core cool. On the third day, water injection failed in Unit 3, and seawater injection began. On the fourth day a hydrogen explosion occurred in the Unit 3 reactor building, and seawater injection started into the Unit 2 reactor.

Over the next several days, operators struggled with small fires, leaks of radioactive gases and water, and uncertainties concerning the spent fuel pool in Unit 4. Eventually the spent fuel pool was confirmed to be filled with water. Water was dropped by helicopters and water cannon to help cool the units.

Approximately 1 week after the event, backup power was restored and water cooling was restored soon thereafter. It will likely be several weeks or months before the full extent of the damage and radioactivity release is known for certain. It is likely that fuel damage occurred in all three units.

New reactor designs that have not been built yet use passive safety features to decrease the probability of failure. Passive systems engage without any human or computer action. For example, in the Westinghouse AP1000 reactor design,[6] a reserve water tank is connected to the reactor core with a one-way valve. The tank is pressurized, but at a lower pressure than the normal operation of the reactor. If there is a loss of coolant, perhaps due to a large pipe rupture, the pressure in the core drops. Once the pressure drops to the backup tank pressure, a one-way valve opens to flood the reactor vessel with water. No human or computer action is required to initiate this backup system, so there is almost no way the safety system could fail. These passive systems will likely be in the next generation of nuclear reactors.

Reactors are also designed to mitigate an accident if the previous systems do not work. The reactor pressure vessel is designed to contain all of the fuel. Beyond that, the building containment is designed to contain any radioactive materials that could be released in the event of a major accident.

When considering the safety of nuclear power, it is helpful to compare historical safety data to other power generation types. Statistical data is taken every year on fatality and injury rates for all industries. This data can be used to determine the total death and injury rate associated with each power option, and it includes all aspects of that energy source—from mining to transportation to actual power generation.

Figure 16 shows a comparison of the fatality and injury rates for coal, oil, natural gas, hydroelectric, and nuclear power.[9] This graph is based on past data from only OECD countries (Organization for Economic Cooperation and Development), which include most of the developed countries in the world. The data is from major accidents that have occurred from 1969 to 1996, so it does not include any minor accidents.

The fatality or injury rate is given as the number of people per 1,000 MWe plant over an average plant life, which is about 40 years. So, for example, the operation of one 1,000 MWe coal power plant will kill 5 people on average over the life of the plant. A majority of these accidents have to do with mining. Likewise there are various death and injury rates from oil, natural gas, and hydroelectric. The majority of accidents with oil and gas have to do with explosion risk, and hydroelectric accidents are dominated by dam ruptures.

Of these five major energy sources, only nuclear power has led to zero injuries and zero fatalities. The only major reactor accident in OECD countries during the time period was the Three Mile Island accident in the United States, which resulted in no deaths or injuries. The Fukushima accident is not included in this data, but this accident did not lead to fatalities. It should also be noted that the Chernobyl disaster was not included in this data since Russia is not an OECD country, but the OECD data is more representative of the risks of U.S. energy generation. All of the other energy sources have much higher risks for accidents, yet nuclear is often perceived to be a more risky power option.

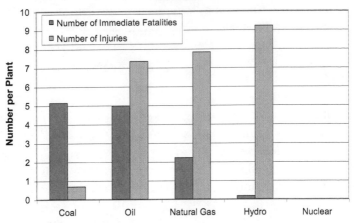

Figure 16: Risk Comparison of Energy Generation[9]

The key point is that all of our most important energy sources in our country have risks associated with them—from coal mining accidents to oil and gas explosions. And by getting a majority of our energy from these sources, we accept those risks. However, based on historical data, nuclear energy is the safest of all these options.

Nuclear Waste

As a result of the splitting of the atom, two fission products are produced for every fission reaction. These fission products cover a large range of elements on the periodic table—some are stable, but many are radioactive with varying half-lives. Other radioactive isotopes are formed due to the neutron bombardment in the reactor core. The fuel cladding is designed to contain all of these isotopes.

After use in the reactor, the fuel is removed and placed into water pools on site until the rods cool down. The past long-term storage plan was to send all of the spent fuel rods directly to a deep geologic burial site at Yucca Mountain, Nevada. However, political opposition has halted the Yucca Mountain Project. Currently alternative waste storage concepts are being examined such as dry cask storage above ground.

The question we should ask ourselves is how nuclear waste affects us on a personal level. The answer is that it does not. We pour billions of tons of CO_2 into the atmosphere with all of our other power options, but nuclear power is the one option that contains all of the waste that it generates. If all of the nuclear waste generated in our country's history were placed in casks and stored on a concrete slab in the desert, it would fit into an area 300 meters by 300 meters by 4 meters tall.[10] Is this really such a problem? Do we really believe that our country is not able to safely manage this amount of material?

The Yucca Mountain Repository was originally designed to safely contain reactor waste for 10,000 years,[11] although there was some debate about whether the life should be extended. I remember going on a tour of the Hoover Dam when I was younger and learning that the dam was designed to last for about 2,000 years, so I have a hard time understanding why we would try to engineer something to last much longer than that. I believe it would be more logical to ensure the safety of the repository for a reasonable amount of time and have faith in our ability to come up with a better solution at some point in the future.

Ultimately, reprocessing of nuclear fuel is the only way to dramatically reduce the volume, heat load, and radiotoxicity of waste. Reprocessing is the chemical separation of spent fuel to remove the dominant long-lived radioactive isotopes. Many of these isotopes can be further fissioned in nuclear reactors to produce additional power. In fact, it is possible to design reactors that can transmute radioactive species into stable isotopes, but a reprocessing plant will be needed to separate the species. It may be possible one day to completely eliminate all long-lived nuclear waste, but that research will not happen if a reprocessing plant is not built.

The difficulty with reprocessing is that it is expensive, and the ability to extract plutonium from spent fuel has been viewed as a proliferation risk. But the extraction of plutonium is the only way that we will be able to significantly extend our uranium resources for a number of centuries into the future.

Environmental Impact

With the exception of hydroelectric plants, nuclear plants are the only source of base load power that generates no greenhouse gases while in operation. The expansion of nuclear power would be the easiest way to immediately cap greenhouse gas emissions from electrical generation. As with any power reactor, a small amount of life cycle emissions are released from the construction of the plant, building materials, and the decommissioning of the plant. A life cycle emissions analysis of nuclear power finds a total of 15 g/kWh of CO_2 released.[12]

There will always be a possibility of radioactivity release into the atmosphere with nuclear power. The probability of this happening is small and can be dealt with through proper engineering. The largest concern right now is that old spent fuel is accumulating at reactor sites in storage pools that were

Nuclear Energy

not designed for long-term use. The waste either needs to be placed in longer-term storage or be reprocessed to dramatically cut down on the total amount.

Economics

Nuclear power has very stringent demands on safety and security that no other source of energy needs to deal with. The Nuclear Regulatory Commission has strict guidelines on siting, licensing, and operating reactors and fuel cycle facilities due to the risks associated with the use of nuclear materials. In addition, with threats of terrorism becoming common, security of nuclear material has become extremely important. Ultimately, these increased demands lead to increased costs for nuclear power that could be enough to keep new plants from being competitive.

The overnight capital cost estimate for building a new advanced light water reactor (ALWR) is about $5,300 per kWe.[13] The cost of electricity for such a new plant is estimated at 11.4 ¢/kWh.[14] It needs to be noted that these costs would be for an additional reactor built at the site of a current nuclear plant—it is common to co-locate 3 or 4 reactors all in one location. This simplifies the construction process if siting studies and environmental impact analyses have already been done for the original plant. The costs for a nuclear plant at a new site are much more uncertain.

Acceptability

After the Chernobyl disaster in Russia and the Three Mile Island accident here in the United States, public support of nuclear power dropped off dramatically. It has taken the many years since then for support to gradually build up again.

The recent nuclear accident in Japan will further test public support for nuclear.

Interestingly, in most areas of the country that have nuclear plants, the local population is more supportive of the technology. Nuclear reactors provide jobs and tax revenue to support the local economy. Those living close to the plants are more aware of how safe the operation of nuclear plants has become. For this reason, additional building of nuclear plants in the coming years will likely be at locations that already have a nuclear reactor. Areas of the country without nuclear plants are likely to be the most opposed to a new plant. It may take longer before a reactor is built at a completely new site.

Conclusion

For the United States to make an immediate impact on greenhouse gas emissions, nuclear plants are an excellent option for satisfying energy demand. The safety of nuclear reactors is proven, and the amount of waste generated is very small when compared to CO_2 emissions from fossil plants. Whether one is for or against nuclear power, it is important to keep in mind what the alternatives are if nuclear plants are not built. The only alternatives for large base load plants are coal and natural gas, both of which will continue to raise CO_2 levels in the atmosphere. Realistically, if our country decides not to build nuclear plants, it will only promote more greenhouse gas emissions.

Chapter 11: Fusion Energy

THE JOKE I LEARNED in my first year as a grad student doing fusion research is that fusion is always 30 years away. When the idea of a fusion reactor was first conceived in the late 1950s, fusion was 30 years away. In the late 1970s and 1980s when the engineering issues of fusion were first investigated, fusion was again 30 years away. And today, fusion is still 30 years away. In the fusion community, we refer to it as the "30-year constant."

The truth of the matter is that fusion turned out to be much more difficult than anyone could have first imagined. There are difficulties with the science, but also difficulties with the engineering. In fact, I believe that it is the engineering issues of creating a fusion power plant that ultimately hold back its development. Although unlikely to be used for power generation in the next 30 years, fusion may be an important power source at some point in the future.

Fusion Technology

Fusion releases energy from the joining of two light atoms. The actual nuclei of the atoms must fuse for energy to be released in the form of heat. Realistically, the way this works is that light atoms like hydrogen isotopes are heated to high temperatures. The outer electron is stripped from the atoms to create a collection of ions—this state is called a plasma. The ionized isotopes are moving very fast at high temperatures and continuously bounce off one another. At these high temperatures, the nuclei of the ions have a good chance of actually fusing together if two come close enough to each other.

Although there are a number of different reactions available for producing fusion energy, the D-T reaction is the most likely to be used in a power reactor. The D-T reaction fuses two different hydrogen isotopes, deuterium (D) and tritium (T). Figure 17 shows the reaction. The deuterium isotope contains one proton and one neutron while the tritium isotope contains one proton and two neutrons. The D-T reaction produces helium (He) and a neutron and releases energy, which translates to a thermal energy that can be extracted.

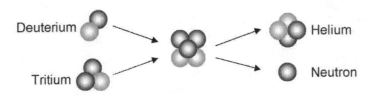

Figure 17: D-T Fusion Reaction[1]

The D-T reaction is the easiest of the different fusion reactions to ignite, although in this case, "easiest" is a relative term. Even after 50 years of research, it is still very difficult to achieve large-scale D-T fusion. Deuterium is plentiful on the planet, but tritium is radioactive and does not occur naturally. However, tritium can be produced in the fusion reactor itself.

The goal of any fusion reactor is to heat up the fuel to the very high energies or densities to increase the probability of fusion. The challenge is to get more fusion energy out than the amount of energy that is required to initially heat the fuel and ignite the reaction. This parameter is called the Q-value and represents the ratio of energy produced to energy input. Energy "breakeven" occurs when the Q-value equals 1, so the amount of fusion energy generated is equal to the amount of energy going into the experiment. The term "ignition" is used to refer to the start of a sustained burn of fusion fuel. When this occurs, the configuration is sometimes referred to as a "burning plasma."

A number of different concepts have been conceived over the years to build a successful fusion reactor. The concepts that are the furthest along can be categorized as magnetic confinement and inertial confinement, though there are some other unique concepts being researched. Unless some major new breakthrough is discovered, it is likely that the working fusion reactor of the future will use either magnetic or inertial confinement.

Magnetic Confinement

In magnetic confinement, a large vacuum chamber is used to contain the fusion gas and minimize impurities. The gas is ionized and heated to high temperatures. Strong magnetic fields are used to confine the charged particles in the reaction chamber. If the magnetic fields were not present, the charged particles would simply hit the chamber wall and lose all their energy. The goal is to confine the ions long enough that a majority of them are able to fuse before they escape or hit the wall. This goal has proven to be much more elusive than was first thought.

Figure 18 shows a picture of the most likely magnetic fusion concept, the tokamak. The fusion chamber is shaped like a doughnut to keep the ions circulating through the chamber without hitting an "end." Strong superconducting magnets surrounding the chamber are required to generate the magnetic fields. A thick liquid coolant also surrounds the fusion chamber to remove the energy generated—this coolant can be used to drive a steam cycle to generate electricity.

Research on magnetic confinement in the past has shown the need for larger and larger reactors to make progress on achieving breakeven and ignition. But large reactors require a lot of research money. For this reason, the International Thermonuclear Experimental Reactor (ITER) project emerged.[2] ITER is an international collaboration to build the first burning plasma tokamak; the United States is a partner. The goals of ITER are to achieve a Q-value greater than or equal to 10 while producing about 500 MWth of fusion energy for about 400 seconds. The reactor will be huge with an overall diameter of 12.4 m and 4 m height. The direct capital cost to build the reactor is expected at $2.75 billion, but the total project cost will be significantly more after accounting for construction time and research time. It is hoped that the research on ITER will lead to new insight into the design of a fusion power plant.

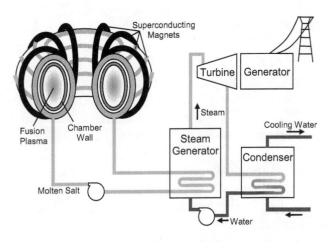

Figure 18: Tokamak Fusion Power Plant

There are many science issues with even making a magnetic confinement experiment produce net fusion power. Instabilities in the plasma or impurities in the system can shut the reaction down, and researchers are working to understand and overcome these issues. However, even if all of these problems are solved, a number of engineering issues exist that make it very difficult to turn fusion into a working power reactor. These engineering issues are examined in a following section.

Inertial Confinement

Inertial confinement is an alternative concept that works much differently than magnetic confinement. A small pellet of fusion fuel is heated and compressed to high energies and densities using some type of external driver. The pellet is usually a cryogenic target with solid deuterium and tritium isotopes. The driver may be either a series of lasers, ion beams, or an x-ray source that quickly heats the fusion pellet. The goal is to ignite the pellet to produce a mini fusion explosion and then absorb the energy released.

Figure 19 shows an example of a proposed laser-driven inertial confinement power plant. A number of laser beams bombard the fusion pellet from different angles. The intense energy pulse heats the outer layer of the fusion pellet very quickly, causing it to explode outward. Subsequently, the inner part of the pellet implodes on itself. The hydrogen isotopes contained within the core of the pellet get heated and compressed to high densities. The fusion reaction is ignited, and then that energy release can be absorbed with a thick liquid blanket surrounding the chamber.

The United States recently built the National Ignition Facility (NIF) at Lawrence Livermore National Laboratory to demonstrate ignition of a fusion pellet using lasers.[3] The NIF design uses 192 lasers with a total energy of 1.8 MJ to heat fusion targets. The chamber containing the target is 10 m in diameter. An alternative inertial confinement fusion facility at Sandia National Laboratories uses an x-ray source to heat pellets and to support the research on the NIF.

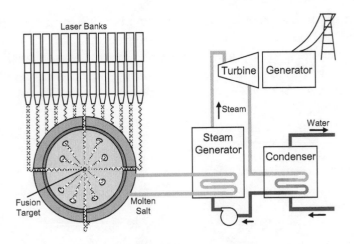

Figure 19: Laser-Driven Inertial Fusion Power Plant

To make inertial confinement into a power reactor, a fusion pellet needs to be ignited again and again. This means that a pellet needs to be dropped into the chamber repeatedly and the driver must be able to fire repeatedly. In effect, inertial confinement reactors ignite mini fusion explosions repeatedly and then remove the energy generated.

The key science challenge with inertial confinement is to produce more power from fusion than is used by the driver to heat up the target. The target needs to be heated enough to start the fusion reaction and initiate burn of a good fraction of the fusion fuel in the target. Many of the engineering challenges surround the high cost of the driver and making a driver that can be fired repeatedly.

Engineering Challenges

It may be obvious to ask why so much work has been done on fusion only to be in a position where it still seems so distant. It is not that the scientific community cannot make

Fusion Energy

fusion work—if the fusion community is given enough money, they probably can build a working fusion power plant. The problem has more to do with the engineering challenges of turning fusion into an economical power reactor.[4,5]

Any D-T fusion power plant, no matter what the configuration, will suffer from the same four engineering challenges:

1. The effect of the intense neutron flux on materials
2. The design of thick liquid blankets for capturing the energy
3. The breeding and containment of tritium, which will be a radiological risk
4. The requirement of expensive drivers

All of these issues can be addressed, but all lead to a technical complexity that considerably drives up the cost of a fusion reactor. To get a better idea of how these issues will affect the economics of fusion, it helps to have in mind an existing nuclear power reactor for comparison.

The D-T fusion reaction releases most of its energy in the form of a high energy neutron, so a fusion reactor will produce a very high neutron flux—much higher than the flux within a nuclear fission reactor. When neutrons bombard materials, they cause radiation damage and activation that limits the lifetime of structural components. This has always been a challenge for fusion in that either components need to be swapped out periodically or very large chamber designs are required (to spread out the neutron damage). Either option leads to higher costs.

The high-energy neutrons require meter-thick liquid coolants to slow down and remove all the energy from the neutrons. These coolants are more exotic materials like molten lead-lithium or molten salts made up of lithium, beryllium, and fluorine. There will be a large degree of technical complexity with designing these thick blankets and pumping the coolant around a heat cycle. Again, the result is an expensive design.

The thick liquid coolant must contain lithium to sustain the fusion reaction. Although deuterium exists in nature, the tritium is radioactive and does not exist naturally. However, tritium can be created in a fusion reactor using lithium. When neutrons bombard lithium, a nuclear reaction occurs to produce tritium. This is one of the most unique aspects of fusion in that it is possible to produce just as much tritium as needed to sustain the fusion reaction. But this also leads to complexities. The systems required to extract the tritium from the coolant are not trivial. Also, tritium is very mobile and can leak easily though structural materials. It will be challenging to keep the tritium leakage at the plant to within regulated limits.

Finally, the driver for any fusion concept is also expensive and technically complex. Magnetic confinement requires large superconducting magnets that are pricey. Inertial confinement requires huge banks of lasers, ion beams, or capacitors for x-ray generation. This requirement really hurts the economics of fusion.

For all of these reasons, fusion will likely be significantly more expensive than other energy alternatives for many decades. It is possible that external factors, like rising gas/oil prices, carbon taxes for coal, or resistance to nuclear fission plants could make fusion more desirable in the future. However, some of the renewable energy options could easily be cheaper than fusion. I do still believe that fusion is a technology that we must research because it may be a much needed energy source at some point in the future, but the research needs to focus on cost-saving techniques.

Development Path

As discussed before, a key difficulty with fusion is that very large and costly machines are needed to be able to reach ignition and breakeven. Fusion is not like other power technologies where the concepts can be tested and developed at a small scale with modest funding. The machines like NIF and ITER are multibillion-dollar projects, and further advances will only call for larger facilities. They must be funded entirely by the government, and there currently is little promise of leading eventually to an economical power plant.

A common theme in this book is that for energy technologies to be able to compete in the electrical generation market there must be a clear development path that gets industry involved to optimize the technology. Fusion does not have this type of development path—currently the only way to reach the fusion goal will be to continue taking multiple billion-dollar leaps of faith.

Fusion will likely need either a major breakthrough or a dramatic change in the development path to reach a more economical reactor. The kind of breakthrough that could make a huge difference would be the ability to reach breakeven with a small experiment. A change in the development path would involve addressing the four engineering issues of fusion in smaller, more economical experiments and facilities. Otherwise, fusion may never be able to get past the 30-year constant.

Environmental Impact

It is a common misconception that fusion is a completely clean form of nuclear power. The advantage of fusion is that the actual by-product of the reaction is helium, which is nonradioactive. Fusion does not produce long-lived nuclear waste like fission does. On the other hand, the strong neutron flux will irradiate reactor components, so components may still have to be disposed of as a high-level or low-level waste. It is possible to use materials that do not create long-lived radioactive isotopes, but these materials are still exotic and only in the research and development stage.

The actual coolant loop will be highly radioactive during the running of the reactor. Materials in the coolant will activate, and the presence of tritium will be a radiological risk. A coolant leak in a fusion reactor could cause quite a mess locally at the reactor site. Materials like lead or beryllium used in the coolant also have toxicity issues. Such a leak could release radioactive materials into the environment if not contained properly.

There are still a lot of uncertainties about the circulation of high amounts of tritium in the plant. Even during normal operation some tritium will leak into the environment. Tritium is a radiological concern because it is highly radioactive and can easily be absorbed in the body (since chemically it acts like hydrogen). The environmental concerns of tritium may be a significant hurdle to building a fusion reactor. Overall, since fusion is still in the research mode, the full environmental impact of a fusion reactor has not been investigated in detail.

Economics

Construction costs for fusion power plants have been estimated, though these estimates are still very immature given the fact that a demo fusion power plant has not been built yet. The most mature technology and most likely (as of now) to be developed into a power plant are variations of magnetic confinement. The ARIES studies are a set of economic analyses of different fusion concepts and provide the basis for the numbers presented here. All of these analyses are theoretical studies and not based on actual machines.

The ARIES-RS project is a reversed shear tokamak that is a variation of magnetic confinement. Total capital costs have been estimated at $4,230 per kWe.[6] The levelized cost of electricity was estimated at 7.6 ¢/kWh. The ARIES-ST is a spherical tokamak that is another variation of magnetic confinement. Total capital costs were estimated at $4,480 per kWe.[7] The levelized cost of electricity was estimated at 7.9 ¢/kWh. A more advanced tokamak, the ARIES-AT, was designed to bring down costs. This design study found a total cost of $2,844 per kWe and cost of electricity at 4.8 ¢/kWh.[8] However, there are a great number of assumptions worked into all of these studies that will depend strongly on how well the research progresses in the future.

Utilities will not be able to build fusion power plants anytime soon, so until the research gets further along, there will not be too much confidence in these cost numbers. The results of the ITER and NIF experiments will be crucial for determining how well these technologies can work as a power plant.

Acceptability

Since fusion reactors have not been built, other than for research purposes, it is difficult to say how well they will be accepted. For now, fusion is still viewed as the ultimate clean energy source, and most people would probably be very supportive of a reactor. The radiological issues addressed above will need to be thoroughly dealt with for a future fusion plant to continue to have support. If tritium leakage becomes a problem in a real site, the concerns over tritium could become just as great as the current concerns about spent nuclear fuel waste.

Conclusion

Although fusion is a research area that I am personally passionate about, the science and engineering hurdles are extremely challenging. Realistically, fusion will probably not play a role in our energy demands in the next 30 years unless some major new breakthrough occurs. It is likely that many of the other clean energy alternatives described in this book will reach economical competitiveness well before fusion. Yet fusion is a huge potential source of energy that cannot be ignored. A strong research program that emphasizes novel approaches and that addresses the engineering issues could help to make fusion power plants a reality.

Part IV

Fossil Energy

THE FINAL PIECE of the puzzle is what to do about fossil fuels. Fossil fuels dominate our energy use—coal and natural gas will continue to be important energy resources in the future. Our greatest challenge in the next few years will be to pass legislation to reduce carbon emissions from these power plants. The final two chapters discuss both coal and natural gas with an emphasis on the technologies that can be used to remove CO_2 from the exhaust. There is no other single technology that will have the most prevalent and dramatic impact on reaching our environmental goals.

Chapter 12: Coal Energy

COAL POWER GENERATION is the largest source of electricity in the United States (and in the world). About 50% of our electricity comes from coal. The proven reserves in North America are believed to be between 210-260 billion tons,[1,2] which could provide enough energy to satisfy our current total electrical demand for 340 years. But coal produces more carbon dioxide than any other power option, so continued reliance on it will have lasting environmental effects.

The greatest challenge for coal power in the next 30 years will be to drastically decrease carbon emissions while keeping the cost of electricity competitive. Carbon sequestration can accomplish this goal, but it will be expensive at first and result in a considerable lowering of the net efficiency of the plant. Strong government support (in the form of regulations and research dollars) will be required to push for carbon sequestration. These steps need to start now if we are to have any hope of achieving a near-zero emission plant in 30 years.

Environmentalists in the past have suggested carbon taxes as a way to encourage carbon sequestration technologies, but such a step could have economic consequences we may not be ready for. We do not want to add a tax that would push coal power generation out of economic competitiveness. There needs to be a gradual development path to reach near-zero emission plants. Government regulation is the only way this will occur, but it should occur in small steps to give the industry time to optimize new technologies. It will only be through wide scale industrial use that costs for these technologies will become manageable.

It has been projected that coal consumption in China could double in the next 30 years.[3] There is no guarantee that other countries will have our same regulations on CO_2—all the more reason that near-zero emission plants must be economical in order for the entire planet to start cutting greenhouse gas emissions.

Coal Power Generation Technology

The traditional coal power plant has evolved over a century but is still relatively simple. Coal combustion takes place in a boiler, and water pipes surrounding the boiler use the heat to generate steam. The high-pressure steam drives a turbine to produce electricity. Almost all of the advances in coal-burning technology have focused on increasing the efficiency of operation or cleaning up the pollution. The following sections describe some of the different coal plant configurations along with pollution-mitigating technologies.

Pulverized Coal Combustion

Pulverized coal combustion (PCC) plants are the most common throughout the country. Pulverized coal is injected along with heated air into a boiler where combustion occurs, typically at temperatures around 1,500 °C. The heat of combustion is used to boil water in pipes lining the boiler—the superheated steam passes through a turbine to produce power. Figure 20 shows a simplified plant diagram for a PCC plant.

The combustion product gases exiting the boiler include carbon dioxide (CO_2), nitrous oxides (NO_x), sulfur dioxide (SO_2), and particulate matter (ash). Extensive regulations exist to limit the emissions of all but the CO_2. The exhaust is first treated to remove most of the particulates using either an electrostatic precipitator or bag filters. SO_2 is removed using flue gas desulfurization (scrubbers). These pollution-mitigating techniques are explained in further detail in later sections.

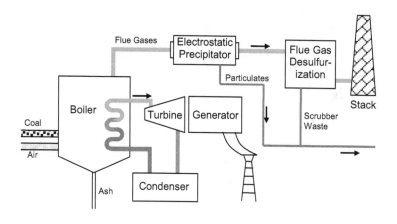

Figure 20: Pulverized Coal Combustion Plant

Since coal power plants have been around for a long time, these systems are well optimized today. Older plants may have conversion efficiencies around 35-40%, but PCC plants built recently reach efficiencies around 43-45%.[4] Higher efficiencies can be achieved by increasing the temperature of the steam and by using higher temperature materials. Nickel-based superalloys that can operate at very high temperatures are being considered for development in the United States and other areas of the world. These new materials could allow for efficiencies around 50-55% to be reached.[4]

Fluidized Bed Combustion (FBC)

A variation of pulverized coal combustion is the fluidized bed combustion (FBC) concept. Crushed coal is fed into a floating bed of ash and coal particles suspended in a stream of preheated air traveling upward through the combustion chamber. The combustion process proceeds in this bed of particles and gas. The heat is used to boil water to drive a steam turbine, but the hot, pressurized exhaust gases can also be used to drive a gas turbine. Whenever a power plant uses two different turbines like this, it is referred to as a combined cycle (see Figure 21). A combined cycle allows the plant to increase the overall electrical efficiency.

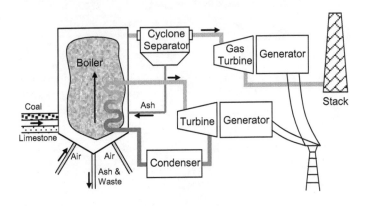

Figure 21: Combined Cycle Circulating Fluidized Bed Combustion Plant

The combustion process proceeds at lower temperatures to minimize the formation of NO_x. Lime or dolomite is fed directly into the fluidized bed to limit SO_2 emission without needing a costly scrubber unit at the back end. The hot gases and bed particles move on to a cyclone or particle separator where the solid material is separated and returned to the fluidized bed. Ash is bled off from the chamber to keep the

Coal Energy

levels in the chamber constant. Electrostatic precipitators or bag filters can be used to limit particulate emissions.

The efficiencies of FBC are about the same as PCC, but the advantages are in pollution control. The feeding of lime directly into the bed allows for reduction in SO_2 emissions without requiring a separate system. NO_x can be reduced by a factor of almost 5 as compared to PCC plants.[4] FBC plants can operate with just about any quality of coal while still achieving good emissions control.

Integrated Gasification Combined Cycles (IGCC)

The cleanest way to burn coal is with the integrated gasification combined cycle (IGCC) plant. In this boiler, pulverized coal is injected with oxygen and steam under pressure to produce a fuel gas (carbon monoxide and hydrogen), also called syngas. Precise control of the oxygen content helps to control the production of pollutants. The fuel gas is then cleaned of particulates and impurities. Sulfur impurities are removed as sulfur dioxide or hydrogen sulfide (useful byproducts). Very little nitrous oxides are formed due to low oxygen levels—ammonia is created instead, a useful byproduct that can be removed from the gases.

The cleaned gas is then burned as fuel in a combustion turbine to produce power. Additional energy can also be extracted from the gasifier and/or from the exhaust gases to drive a steam turbine. Figure 22 shows a typical schematic.

IGCC plants can be complex systems requiring long startup times. They do achieve efficiencies in the 45% range and may reach as high as 56-60% in the future.[4,5] Their main advantage is the ability to achieve very stringent emissions standards on all pollutants that are of concern. For this reason, IGCC plants may prove to be useful in the future.

The cost of IGCC plants is higher than other coal technologies. Cheap natural gas prices in the 1990s led to increased use of gas turbines, which in turn limited IGCC development. As environmental regulations become stricter on coal power plants, IGCC may start proving to be more economical than other options.

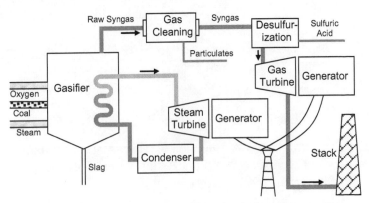

Figure 22: Integrated Gasification Combined Cycle Plant

Environmental Impact

Without emissions regulations, the burning of coal is a dirty process. Pollutants like SO_2 and NO_x cause acid rain, which can have devastating effects on the environment and structures. NO_x and particulates can cause smog, which leads to health problems. Of more recent concern is control of mercury emissions because of the toxic effect of heavy metals in the body. Existing regulations have helped clean up coal considerably, but the large amount of CO_2 released is still one of the major concerns. The Department of Energy has an initiative called Vision 21 whose goal is to remove the environmental concerns of fossil fuel use.[6] This program supports research and demonstrations with very stringent emissions goals, including CO_2 removal.

NO_x, SO_2, Particulates & Mercury

Nitrous oxides (NO_x) are typically prevented using low NO_x burners. Controlling the introduction of air and the temperature of combustion helps to reduce NO_x by 30-60% in standard plants. IGCC plants typically produce much lower levels of NO_x. For very stringent demands, selective catalytic reduction (SCR) can be used to reduce NO_x by as much as 90% (if regulation requires it).[4,7]

Scrubber technology, or flue gas desulfurization (FGD), is used to remove SO_2 from plant exhaust in conventional power plants. (In the fluidized bed concept, sulfur is removed directly in the bed by adding limestone, which simplifies the process.) Wet scrubbers, the most common technique, use a lime or limestone slurry to absorb sulfur from the exhaust. Scrubbers have come down considerably in cost in the past 30 years—almost a factor of 4 reduction. Today, scrubbers add an additional $80-$120 per kWe to the cost of a plant.[7] Both traditional PCC plants with scrubber technology and fluidized bed plants typically reduce SO_2 emissions by 90-95%, but IGCC plants can reach 99% removal rates.[4,7]

Particulate emission is removed with the use of electrostatic precipitators and/or bag filters. Electrostatic precipitators consist of parallel plate electrodes through which the flue gas passes. A voltage is applied across the plates high enough to charge the particulate matter—the ash and dust are collected on the plates and periodically removed. Bag filters simply use cloth to filter out dust particles. The electrostatic precipitators are more cost effective for larger plants, costing from $40-$60 per kWe. Bag filters are more effective at removing very small particles and cost from $50-$70 per kWe.[7] Through the use of either option, 99-99.7% of the particulate matter in exhaust can be removed.[7] IGCC plants produce almost negligible amounts of particulates due to their unique operation.

More recently, in the United States, mercury emissions are being targeted as part of the Clear Skies Initiative, which calls for emission cuts of 70% by 2018. Mercury emissions vary widely depending on the coal. Different technologies are currently being investigated for mercury control, but it is likely that the scrubber technology used to remove SO_2 should be useful for removing most mercury as well. If additional measures are required, activated carbon injection may also be used. Future goals suggest 90% mercury removal as a goal for advanced plants.[3]

CO_2 Removal

The real focus on new coal technologies in the near term needs to be on limiting CO_2 emissions. Coal power plants produce the most CO_2 of any power generation option. Life cycle assessments find a release rate of 974 g/kWh.[8] The majority of this release is from the actual burning of coal, but this number also includes life cycle emissions from plant construction, decommissioning, and extraction and transportation of coal. Carbon sequestration technologies can reduce CO_2 emission, but the technology is immature and expensive. It also consumes a considerable amount of power, which would reduce the net power production from a plant. Removing and then sequestering CO_2 in the ground in an economical manner will be the biggest challenge for coal power in the coming decades.

One method for the removal of CO_2 involves scrubbing the exhaust with chemical agents or with solids that absorb CO_2. Then in a regenerative step, the agents or solids are heated to drive off the purified CO_2 so that it may be compressed and liquefied for sequestering in the ground. Unfortunately, the heat required for regeneration and energy required for compression is considerable and results in an overall decrease in the net efficiency of the power plant by 10-12 percentage points.[4] This efficiency drop would hurt the economics of coal power plants significantly. Preliminary work suggests that CO_2 removal may be most efficiently accomplished with IGCC plants, with a loss of 6-8 percentage points on efficiency and possibly only a 4% reduction in the future.[4,5] This is one more reason why IGCC plants will likely gain in popularity in the coming years.

Other methods have been studied as well. Membrane technology can be used to separate out the CO_2. Or the exhaust gases can be cooled to condense out water, and then the leftover CO_2 can be compressed for sequestration. Another method that may be better suited for sequestering is to react hot CO_2 with magnesium or calcium silicate to form carbonates that can be stored in the ground. These technologies are all in early stages of development, but could allow for regeneration stages using less energy.

The FutureGen project in the United States aims to build a large-scale research power plant capable of CO_2 sequestration and hydrogen production. The goal is to modify an existing plant to remove 90% of the CO_2 released. The goals are challenging and include keeping the overall levelized cost of electricity to less than 10% higher than non-sequestration plants.[4]

Carbon Sequestration

After the CO_2 is removed, it then needs to be sequestered in the ground. In certain areas of the country, CO_2 can be pumped into caverns or wells. Generally, CO_2 would be injected into the ground in a supercritical form. Oil fields, gas fields, saline formations, and unmineable coal seams may all be considered. Since we know a great deal about the geology of hydrocarbon formations, these sites may be ideal for initial studies. As described above, other schemes investigate chemical reactions to sequester the carbon in a solid form that can then be disposed of.

There may be a large potential for CO_2 sequestration throughout the world. Studies have examined the cost of transportation and disposal of CO_2 in the ground and found that the transportation and pumping of the CO_2 into the ground are relatively small costs compared to the capture technology.[5] Therefore, research should be focused on decreasing the cost and energy requirements of the capture process.

As carbon sequestration projects have moved forward, so to have concerns about the safety of CO_2 storage underground. Because the idea is relatively new, data will be limited until demonstration projects are completed. The potential for leakage not only partially negates the benefit, but it can also be a health threat if leakage occurred in a more dramatic event. Monitoring and safety will be a key measure of these projects in the future.

Carbon Regulations

Different ways of regulating CO_2 emissions have been proposed in the past. A carbon tax is one method that would simply involve a tax on all producers of CO_2. Such a tax could help push for carbon sequestration technologies, but its implementation could be problematic. A carbon tax could have far-reaching, unintended economic consequences. If a carbon tax is applied, the cost of coal power generation will immediately increase, and who do you think is going to pay for it? The cost increase will be passed on to us, the consumers.

Carbon caps are another method to limit emissions that simply place a cap on the amount of CO_2 that a country can emit. Utilities and industries are given an allowance, and they need to do what they can to keep below that allowance. However, carbon caps provide industry with too many ways to circumvent the problem, allowing companies to trade carbon emissions as part of this system. Carbon caps could easily drive domestic industries out of business as other countries without such caps will be able to make products more cheaply. It makes no sense to limit carbon emissions in the United States if it only promotes the growth of dirtier industries in another part of the world. Caps also can make electricity prices soar if power plants must limit output to stay within their allowance.

Regulations need to be designed to prevent such unintended consequences. The carbon tax and the cap and trade system allow for too many of these types of problems to occur. Ultimately, these systems will limit growth and force more reliance on overseas companies. The point is not to make electricity and the cost of products more expensive. The point should be to force the addition of carbon-sequestering technologies.

It may make more sense to place regulatory limits on emissions that the utilities must satisfy with carbon-sequestering technologies. Realistically this will only occur in small increments to allow the industry time to develop the technology. For example, legislation could start with a 20% cut in CO_2 emissions, but the regulations cannot allow the plant to simply reduce power production by 20%. This would force coal power plants to use carbon sequestration technologies, but only on a small scale to start. It should not change the costs of electricity much, and it allows the technologies to be optimized on the commercial scale (similar to how scrubber technology has improved over the years). Costs will begin to come down, which will help when the regulatory limits get tighter.

This type of regulation would need to start now to have any hope of achieving a near-zero emission plant in 30 years. The point is that industry cannot avoid developing carbon-sequestering technologies by simply shutting down for parts of a day. We need industry to adopt these technologies on a large scale to begin to bring down costs.

Economics

Because coal power is so well established, a number of economic assessments of different plants are available. The technical risks of basic coal power generation are minimal. The financial risks are in public acceptability and increasing future regulations of emissions control. The type of coal to be burned needs to be taken into account, but coal supply in this country is mostly stable, so prices should be stable for some time into the future.

The capital cost for conventional PCC and FBC plants following current emissions controls averages $2,800-$3,200 per kWe installed, while IGCC plants may range from $3,200-$3,600 per kWe.[9] The levelized cost of electricity for these plants is likely to range from 9.5-10.9 ¢/kWh.[10]

The costs to build a coal plant with carbon sequestration are considerably higher, but these costs have a lot of room to decrease as the technology matures. The capital cost estimates for PCC w/ sequestration average $4,500-$5,100 per kWe, while IGCC plants with sequestration could cost around $5,300 per kWe.[9] The average efficiency for these plants is expected to be lower by about 6-12 percentage points. The levelized cost of electricity is expected to be about 13.6 ¢/kWh.[10]

It is clear that carbon sequestration technologies as they currently stand will hurt the economics of coal, though they may still be competitive with other clean energy options. It will take substantial research and demonstration to bring down the costs to more palatable levels. In the near term, small reductions in CO_2 emissions (on the order of 20% to start) will not affect economics too much. It is likely to be a long road of small steps in additional regulations and technology development before full carbon sequestration can become an economic reality.

Acceptability

Public acceptability of coal power is one of the major weaknesses of this energy option. Although coal power plants have considerably cleaned up the amount of NO_x, SO_2, and particulates, the high production of CO_2 still causes people to view it as a dirty power option. A recent CBS News/New York Times poll found 41% of respondents in support of building more coal plants and 51% against.[11]

Since coal power plants are typically built close to the source of the coal (which usually means they are close to populations that are dependent on the mines or power plant for local jobs), these areas tend to be more accepting of the technology. The best hope of making coal more acceptable is to work to eliminate CO_2 emissions with carbon sequestration. The previous poll found that 69% of respondents would support cleaner coal plants that reduce emissions, even if they cost more.

Conclusion

As one of our most abundant natural resources, we will continue to rely on coal for electricity for many decades in the future. Modern coal plants are economically competitive, but the production of CO_2 is coal's biggest downfall. Our greatest challenge in the next few decades will be to push for gradual carbon regulations to develop carbon-sequestration technologies on the industrial scale.

Carbon-sequestration technologies are the best chance of transforming coal into a clean energy source, but it will take years of government regulation and research support to reach near-zero emission plants that are economically competitive. Industry will fight carbon regulations with all their power. However, change will eventually be coming, and those industries that develop the sequestration technologies early will be in a much better position when the regulations do come.

Chapter 13: Natural Gas Energy

NATURAL GAS POWER GENERATION has seen a rapid expansion in the past couple of decades due to a number of reasons. Today's gas turbines have high efficiencies, and emissions from natural gas are cleaner than from coal. In addition, natural gas turbines are cheap to install. With so much uncertainty about building large base load power plants, natural gas made the most sense to utilities looking to expand power generation or to add peaking capacity during high demand months. In the United States, power generation accounts for only about 31% of natural gas use, whereas the rest is used predominantly for residential heating and industrial processes.[1]

One of the problems of relying on natural gas for energy production is the volatility of the price of natural gas. Since natural gas turbines are cheap to install, a majority of the cost of electricity depends on the price of the fuel. A volatile natural gas market can lead to volatile electricity prices.

Over the past 30 years, the total U.S. extraction of natural gas has not changed much. However, recent discoveries offshore and new drilling technologies have significantly increased our domestic supply projections just over the past few years.[2] As a result, natural gas prices have come down, and natural gas appears to be promising for expanded use in the coming years. Yet carbon regulations will be a concern for increased natural gas use, and we should consider whether natural gas is best used for baseload power.

Gas Turbine Technology

The basic gas turbine design consists of a compressor, combustion chamber, and gas turbine (see Figure 23). Modern turbines have the compressor and turbine on the same shaft for increased efficiency. Intake air enters the compressor, where it is compressed to 15-20 times atmospheric pressure. Then the air is mixed with natural gas in the combustion chamber where it burns at temperatures up to 1,400 °C. The pressure increase from the heated gas drives the turbine to produce power.

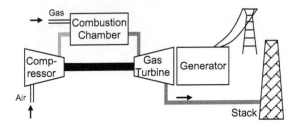

Figure 23: Simple Gas Turbine

Typical gas turbines for power production may produce up to 250-300 MWe. Efficiencies as high as 38% can be achieved for a simple turbine, though future development will push up the efficiency as high-temperature materials are used. A number of different performance enhancements can be used with a gas turbine. Reheating, intercooling, mass injection, and recuperation are all techniques to boost the efficiency by a few more percent.[3]

Natural Gas Energy

The most efficient and cost-effective natural gas plant today is the natural gas combined cycle (NGCC) plant. A combined cycle plant uses both a gas turbine and a steam turbine to maximize power extraction. Figure 24 shows a schematic. The difference from a stand-alone gas turbine is that additional energy is extracted from the hot exhaust leaving the turbine. The hot gas is used to boil water in a steam generator, which in turn is used to drive a steam turbine to produce electricity.

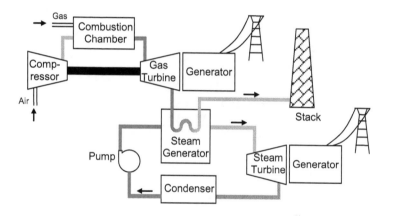

Figure 24: Natural Gas Combined Cycle Plant

Combined cycle plants can reach efficiencies near 57%, and future development will push them above 60%.[3] These high efficiencies have helped to keep natural gas power plants very economical even when gas prices were high in recent years. The higher efficiency results in less pollution per amount of electricity generated. Combined cycle technology is the most efficient production of power available today, and the technology is incredibly optimized.

The greatest advantage of natural gas is the cleaner output (as compared to coal). Natural gas plants do not require all of the expensive pollution controls that coal plants require for particulate and sulfur removal, making installation costs much cheaper. On the other hand, natural gas does still produce significant amounts of CO_2 which must be addressed with carbon sequestration technologies.

U.S. Natural Gas Reserves

The proven reserves of natural gas in the United States are estimated around 238 trillion cubic feet, but total technically recoverable resources are estimated around 2,100 trillion cubic feet.[4] The total resources are enough natural gas to satisfy U.S. gas demand for about 90 years. The recent advent of horizontal drilling and hydraulic fracturing technologies have put many of these resources within our reach, so our country is suddenly in a position of having a wealth of natural gas to rely on.

Directional drilling provides the capability to drill down and then horizontally to access more gas resources from one well head. Hydraulic fracturing involves the injection of water and chemicals into shale formations to recover gas or other fossil fuels. The perfection of these technologies in the last couple decades has allowed for the economic recovery of shale gas, which has dramatically changed the outlook for natural gas in the United States.

Like many in the country, I have a natural gas furnace in my home, and I like the way gas heats up a house. Residential or industrial heating is the most efficient use of gas since heat is not lost in a thermodynamic cycle. The use of gas for residential heating may be 80-90% efficient, so from an efficiency standpoint, it makes sense to reserve this resource for direct heating needs.

An advantage of gas turbines is their ability to come on-line quickly, which makes them ideal as peaking turbines to produce extra power on hot summer days when air-conditioning is stressing the grid. Gas turbines are also ideal for making up lost demand for renewable technologies. For

Natural Gas Energy

example, it could be useful to have backup gas turbines available in conjunction with a wind farm to come on-line in times of low wind speeds.

Environmental Impact

Compared to coal power, natural gas produces much less pollution, and the production of CO_2 is reduced. As with coal, NO_x is generated in the combustion process. The amount depends on the temperature of combustion—as technology continues to push for high temperatures (for increased efficiency), NO_x production will become more of a problem. Low NO_x burners control the mixture of gas and air to limit NO_x and carbon monoxide emissions. Newer gas turbines can typically keep the emissions less than 10-25 parts per million (ppm). If requirements that are more stringent are in place, selective catalytic reduction (SCR) can be used to reduce emissions to 5 ppm.[3]

A key advantage of natural gas is that SO_2 emissions are negligible, and there is no particulate emission. So natural gas plants do not require all the additional expensive emission controls that modern coal power plants need.

Natural gas combustion produces about half of the CO_2 that coal power plants emit for the same unit of electricity delivered. Many countries around the world have used gas to replace coal power to help meet the goals of the Kyoto Protocol. Life cycle greenhouse gas emissions have been estimated at about 500 g/kWh CO_2 equivalent.[5] Actual life cycle emissions are 440 g/kWh of CO_2 and 2.8 g/kWh of CH_4. Since methane is a much stronger greenhouse gas than CO_2, it leads to a higher "CO_2 equivalent" release.

The same carbon-sequestration technologies that can be applied to coal exhaust can also be used to clean up natural gas exhaust. Since there is less CO_2 produced per unit of energy, it will be easier to remove carbon as compared to coal plants. However, it will still be costly and lead to significant drops in overall efficiency. To be fair, any future regulation of carbon emissions will need to target natural gas as well as coal. Also, if only coal were targeted, it would lead only to increased use of natural gas.

The question with carbon sequestration of natural gas is where to draw the line. Gas is used for many different purposes including residential, commercial, and industrial heating as well as for power generation—should emissions for all uses be regulated? Since we have already established that natural gas is much more efficiently used for heating, it probably makes more sense to regulate only carbon emissions for gas turbines that are used to generate electricity. As with coal, these regulations should be such that they only allow for the addition of carbon sequestering to reduce emissions.

I can easily see that such a step could generate a great deal of controversy in the future. An industry that uses gas turbines to produce electricity will have to add expensive carbon-sequestering technologies, while an industry that uses natural gas only for heating will not have to make any changes. But the whole point is to discourage the use of gas for electricity when it can be used much more efficiently for heating.

Economics

The capital cost of new combined cycle natural gas plants is estimated from $970-$1,000 per kWe.[6] This is very low compared to other sources of electricity, and indicates how optimized the technology is today. The levelized cost of electricity is expected to be around 6.3-6.6 ¢/kWh for a combined cycle plant at the current price of gas.[7,8,9,10]

Estimates for near-zero emission natural gas combined cycle plants are around $2,000 per kWe, but as with coal, there is a great deal of room for prices to drop as the technology matures.[6] The cost of electricity is expected to be around 8.9 ¢/kWh.[7] Overall efficiency of near-zero emission plants will likely drop by about 8-12 percentage points, though it is hoped the efficiency losses will be lessened as the technology is optimized in the future.[8,9]

Acceptability

Natural gas power plants are a well-accepted power generation option today. Gas has been viewed as a good alternative to coal since it does not produce the additional pollutants like coal and emits half the CO_2. Most natural gas plants are smaller, so they are nonintrusive on land and individually do not produce much pollution.

Natural gas still contributes significantly to climate change, so its use may see more resistance in the future. If natural gas prices stay reasonable, it will continue to be an attractive energy option. Similar to coal, near-zero emission technologies will help to make gas an even more acceptable solution in the future.

Conclusion

With advances in drilling technologies, the United States can continue to rely on natural gas for several decades. With low capital costs and reasonable gas prices, it is likely that its use will expand. Natural gas is more efficiently used for industrial and residential heating, and as a way to enable the development of more renewable technologies. It may be a better use of our resources to limit the use of gas for base load electricity generation and instead save it for heating and as a way to stabilize the grid in periods of high demand.

Zero emission natural gas plants use the same technology as zero emission coal plants, so these technologies should be pushed so that they are ready for future use. If future regulatory changes call for reduced carbon emissions from coal plants, it would be only fair to ask for the same reduction from natural gas plants.

Conclusion

BY THIS POINT, all of the alternatives to the major producers of energy have been discussed. Part I examined the advanced transportation options to reduce dependence on oil and reduce emissions. Part II examined the renewable energy options and which technologies have a realistic chance of being built on the large scale. Part III went through the merits and challenges of nuclear energy. Finally, Part IV discussed the fossil fuels and ways to reduce emissions. So where should our country go from here?

Chapter 14: The Energy Construct

THE ENERGY CONSTRUCT contains all the elements that are needed to achieve a clean, domestic, and economical energy future (see Figure 25). This figure summarizes the lessons learned from this research to provide a roadmap for the future of energy in our country. Both near term and long term elements are included.

Figure 25: The Energy Construct

Four technology paths have been identified as the most likely to make significant changes in the way we generate and use energy. The inside ring of the construct are those technologies that can be feasibly built now, and that could make a dramatic difference over the next 20 years. The outer ring is more uncertain, but represents those technologies that could make a difference in the 2030-2050 time period.

Of the fossil fuels, both coal and natural gas are in abundance in the United States. Efficient natural gas plants can help to reduce emissions in the near term. But the greatest challenge will be to push for carbon-sequestration technologies now to force industry to optimize the systems. Only incremental reductions in carbon emissions will be likely in the next 20 years, but the long-term vision is to achieve zero emission coal and natural gas plants.

Of all the renewable energy options, only wind energy is economical on the large scale for the time being. This technology is an excellent way to achieve emission-free power—the capital costs are reasonable and the modularity of wind turbines is desirable for utilities looking to expand their renewable energy portfolio. Geothermal can be economical but is limited by suitable sites. Solar is not competitive but is being aggressively pushed to reduce costs. Initially these other renewables will depend on subsidies to expand, but it is hoped that in the long-term we can achieve a diverse portfolio of non-subsidized renewable power plants.

New light water nuclear reactors are the only way to immediately increase base load power without greenhouse gas emissions. The safety of these reactors is proven, and electricity can be generated economically. The greatest challenge with nuclear energy will be in removing the political and financial roadblocks associated with this technology. In addition, the country must move forward with a permanent solution to the storage of nuclear waste or move toward reprocessing in the long-term. In the future it is likely that more advanced light water reactors designs (possibly

The Energy Construct

including small modular reactors), will be built that utilize increased reliance on passive safety.

The final portion of the energy construct is not an energy source, but it will be the only way to reduce dependence on foreign oil. Plug-in hybrids will be the first step in reducing oil use. They will be the most effective in reducing carbon emissions and using energy resources efficiently, but the other clean energy options must be developed in parallel for electric vehicles to make a positive environmental impact. In the long term, the plug-in hybrid will lead to the electric vehicle.

Comparison of Energy Options

With all of the data presented throughout this book on the various power generation technologies, it is useful to see a side-by-side comparison. Figure 26 shows how all the generation technologies stack up to one another based on economics, environmental impact, domestic resource potential, public acceptability, and reliability. (The Appendix also compiles the economics and emissions data in a table.)

The top graph shows a comparison of the levelized cost of electricity for new plants. Each bar is plotted as a range to reflect uncertainty in costs—the bottom of each bar represents a low estimate while the top of each bar represents a high estimate. The next graph compares the total life cycle CO_2 emissions from each power source. The plot of resource potential is more of a relative scale but still important in choosing future energy sources. The graph of public acceptability is based on public opinion polls if the data was available, so it represents the percent of Americans

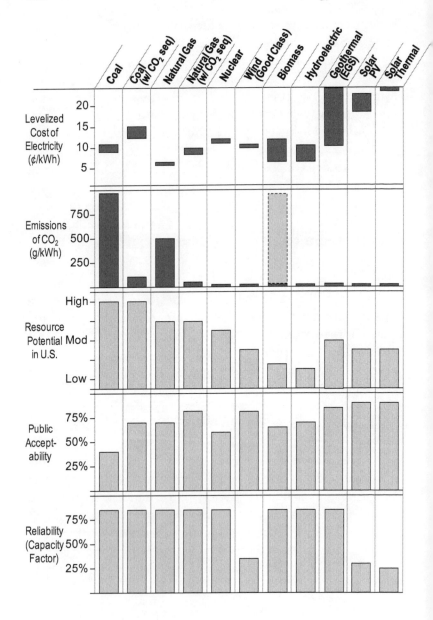

Figure 26: Power Generation Summary Comparison

supportive of expanding that technology. (But keep in mind that polling data can be misleading.) The final graph represents reliability given as average capacity factor.

The energy technologies not included on this graph include oil, ocean energy sources, and fusion. Oil was not included since this is an energy source that we want to move away from. Ocean energy sources and fusion still require quite a bit of research to be able to make accurate cost predictions, but they could be important sources of power in the longer term.

One of the advantages of comparing all of the energy options side by side like this is to easily see the strengths and weaknesses of each option and then decide if solutions to the problems are possible. The following conclusions can be made:

1. The key problem with coal is the high CO_2 emission rate, which in turn is mostly responsible for low public acceptance. Carbon-sequestration technologies can solve this problem with the tradeoff of a higher cost of electricity. However, costs will need to come down for zero emission plants to be competitive.
2. Due to a good resource base and low electricity costs, natural gas looks attractive at the current time. Natural gas produces significant CO_2 emissions, which could also be dealt with through carbon sequestration. Interestingly, natural gas plants with CO_2 sequestration appear to be competitive, but no actual plants have been built yet.
3. Nuclear energy is a good power option in terms of cost, emissions, resource potential, and reliability, but public acceptability is still a hurdle. The high cost for building new plants is also a hurdle even though levelized costs are competitive. It will take both strong public and government support to get new plants licensed and built.

4. Of the renewable energy options, wind has the most potential—areas with good wind resources are economical, but costs will still need to be reduced for areas with lower wind speeds. Reliability is still an issue, but wind turbines could produce as much as 10-20% of the country's power without adversely affecting the electrical grid.
5. Biomass has a couple weaknesses including considerable immediate emissions of CO_2, and a low resource potential. The resource potential was rated low since huge amounts of land are required for large biomass plants. Biomass will be unlikely to expand on a large scale, but smaller-scale uses like waste burners and digesters can make sense.
6. The fundamental problem with hydroelectric power is the low availability of potential sites. This option will be unlikely to expand.
7. Geothermal will only expand through the use of enhanced geothermal systems, but costs are currently too expensive for widespread use. Limited expansion may be seen using old oil and gas wells, but costs will need to come down considerably for geothermal to expand on a large scale.
8. Solar is also a very expensive option with the further problem of limited reliability. Unless costs can come down significantly, solar use may expand on only a small scale in limited areas of the country.

Research Requirements

Based on the conclusions of this book as shown in the previous two figures, a list of the required research has been generated. This list contains the top 10 technologies that will

The Energy Construct

make the most difference in achieving clean energy. The list is prioritized in order of importance.

1. *Carbon Sequestration*: Carbon-removal technologies will be one of the most important advances of engineering in the next few decades. A very strong research and development program should be supported along with strong industrial incentives or regulations.
2. *Battery Technology*: Continued improvements must be made in battery life, size, and cost to help push the plug-in hybrid and electric vehicle development path. Again, this will require a combination of a strong research program along with industrial incentives.
3. *Rapid Charge*: The rapid-charge capability for batteries will be the only way to make a transition from plug-in hybrids to all-electric vehicles. There is still some time to develop the technology, so this should be a strong research program.
4. *Wind Energy*: Improvements will be needed to make wind power economical in areas of lower wind resources, but these improvements will depend on more commercial expansion of wind turbines.
5. *Nuclear Energy*: The challenge with nuclear power is sociopolitical as opposed to technical. Work will be required to remove the barriers to licensing and siting new plants. Government incentives will be required to get new plants built.

The above 5 technologies will contribute to the inner ring of the construct. Both strong research and demonstration programs along with industrial partnerships will be important in achieving these future goals. Of less importance in the near term, but with more potential in the long term, are the following 5 technologies:

6. *Solar*: Solar energy is still very costly, so continued research should focus on cost-saving techniques. Continued government incentives to build large-scale plants in the Southwest area of the country should be continued
7. *Wave Energy Converters* & *Ocean Current Turbines*: Ocean energy technologies have a high potential in the long term due to the higher power density and more reliable power output as compared to wind energy. These technologies should be supported now in a strong research program.
8. *Nuclear Fuel Reprocessing*: Reprocessing technologies are the only way to drastically reduce the amount of high-level nuclear waste in the country and extend uranium resources. The research on reprocessing should be strongly supported now so the technology is ready in the future when it is needed.
9. *Enhanced Geothermal Systems*: In the near term, enhanced geothermal technology can be used to produce geothermal energy from aging oil and gas wells. It will be a challenge to bring down costs to make this option available in all areas of the country.
10. *Fusion*: Fusion will be an important power source at some point in the future, but many more years of research are required. The fusion community should consider alternative concepts that are better able to address the engineering challenges of fusion reactors.

Two technologies that are notably absent from the list are biofuels and hydrogen. Biofuels such as ethanol make no sense as an advanced transportation fuel due to a lack of enough farmland and since they do nothing to help reduce pollution. And the hydrogen fuel cycle will be incredibly wasteful of energy, requiring twice as many new power plants as compared to the use of electric cars.

Climate Change

What effect will these changes have on total CO_2 emissions? In 2010, the United States emitted about 5.4 billion metric tons of CO_2 into the atmosphere. Almost all of those emissions were from oil, coal, and natural gas.

If a combination of plug-in hybrid and electric vehicles dominate the transportation sector by 2030, it would be possible to decrease emissions from oil by a factor of 2. If aggressive carbon regulations require 50% removal of CO_2 in coal and natural gas by 2030, these emissions could be reduced by a factor of 2 as well. More power plants will be required for all of the plug-in hybrids and electric vehicles, but that increase can be balanced with new renewable or nuclear power generation. If all of these options occur, U.S. CO_2 emissions could easily drop to 3-4 billion metric tons per year by 2030, with more reductions beyond.

The Electric Economy

One of the themes of this book is to move our country into an all-electric economy. The use of plug-in hybrids and electric vehicles will require the addition of much more clean power generation, but the use of electricity to power our cars will save us considerable money on fuel costs. Chapter 4 showed how the average American may spend $1,800 per year on gasoline, but costs would be reduced to $450 per year on added electricity if an electric car were used instead. Not only does this allow us to be able to spend more on plug-in hybrids and electric cars, but it also allows us to spend more on clean sources of electricity.

Consider that the average American may spend around $900 per year on residential electricity, heating, and cooling—which means that we spend about twice as much on gasoline as we do on our utility bills. By switching to an electric car, even if clean sources of electricity cost 25% more, we will still come out ahead in the all-electric economy.

What Can You Do?

Many of the conclusions of this book are on the large scale, so you may be left wondering what you can do. Probably the best action you can take to reduce dependence on oil and reduce emissions is to buy a fuel efficient car, hybrid, plug-in hybrid, or electric car (when they are available) when it comes time for your next car purchase. If you are part of a family and concerned about passenger room, consider having one family car and one efficient hybrid. At your next car purchase, take some time to think about the conflicts in the Middle East and our appetite for oil. A desirable all-electric vehicle will still take some time to develop, but the purchase of hybrids and plug-in hybrids will speed up the development path.

Consider becoming more aware of any new power plants being proposed in your city or state. Many new plants are met with a local opposition group. If you are supportive of the technology, consider finding out if there is a local group that supports the project.

Many utilities now offer programs for individuals to invest in renewable energy. For more information, you can try to find out if such a program is available from your utility. Usually it involves pledging to pay more for electricity to support development of renewable sources. This might seem a little odd since a utility cannot actually send you electricity that comes only from a clean energy source, but the idea is that if enough people support paying more for renewable energy, the utilities can justify building more renewable power plants.

Lastly, one of the best things you can do is to support politicians who have sensible energy policies. In this day of high gasoline prices, our politicians should become much more informed on which energy options make the most sense. Our world is going to suffer the consequences if politicians are not able to make changes now.

Final Thoughts

The future that I envision is one that will require a significant investment in energy technologies, but these investments will more than pay for themselves in the long run. The interstate highway system cost the U.S. government millions of dollars when it was established. However, we probably could not begin to calculate how much it has helped our country in providing opportunities and expanding commerce. Our energy future needs this same sort of infrastructure development.

This investment is difficult in our government. The reason has a lot to do with term limits, but also with external factors in the world we live in today. Our government usually makes a major transition every 4 to 8 years, whereas the research and development time for most energy technologies may take 15-20 years. So often projects get funded for a few years and then get killed when the next administration comes in and is not as favorable to that technology. At the same time, the government is always squeezing the budget, and this leads to cancelled or cut back projects. Our government leaders need to have a long-term vision to keep these projects going.

It is easy for a candidate to support technologies like conservation and renewable energy—after all, who does not support these options? But we need leaders with the backbone to make the hard choices. We need leaders who are willing to take a strong stance on reducing CO_2 emissions from coal and gas. We need leaders who are strong enough to support nuclear power. We need leaders who recognize that electric vehicles are the only logical clean transportation option of the future.

In 30 years, I see the highways dominated with electric and hybrid vehicles that have significantly reduced our dependency on oil. I see coal and natural gas plants with "carbon scrubbers" to eliminate most of the CO_2 emissions. I see more nuclear plants designed with passive safety systems and designed to minimize waste. I see many wind farms throughout the country. I hope to start seeing more solar power and geothermal energy along with wave and ocean current converters.

This future is not science fiction; it is within our reach. With vision, public support, and political will, we can achieve our goals of providing energy that is clean, affordable, and domestic. It is time to lead the world in a clean energy revolution.

Appendix

Energy Source	Plant Type	Overnight Capital Cost ($/kWe)	Levelized Cost of Electricity (¢/kWh)	Greenhouse Gas Emission (g/kWh)
Coal	Conventional	2800-3200	9.5	974
	Conventional with CO_2 sequestration	4500-5100	13.6	97
	Gasification	3200-3600	10.9	779
	Gasification with CO_2 sequestration	5300	13.6	78
Natural Gas	Combined Cycle	970-1000	6.3-6.6	500
	Comb. Cycle with CO_2 sequestration	2000	8.9	50
Nuclear	Advanced Light Water Reactor	5300	11.4	15
Wind	Land-based (Good Wind Resources)	2400	10	15
	Off-shore	6000	24	15
Biomass	Direct Burning	3860-7900	6.2-11.6	—
Geothermal	Natural Steam Site	1400-1500	5-8	—
	Binary Plant	2000-2100	5-8	—
	Enhanced Geoth. Systems	5000	10-30	—
Hydroelectric	Conventional	1500-3000	6-10	—
Solar	Solar Trough	4700-5010	31.2	—
	Solar Tower	5140-9090	31.2	30-43
	Solar-Stirling	4000-6800	19.7-44.3	—
	Photovoltaic	4750-6050	21	39

Table 4: Energy Summary

References

Chapter 1

1. "Annual Energy Review 2009," DOE/EIA-0384(2009), Energy Information Administration, available at www.eia.doe.gov (March 2011).
2. "Annual Energy Outlook 2010," DOE/EIA/0383 (2010), Energy Information Administration, available at www.eia.doe.gov (March 2011).
3. "Climate Change 2007," Intergovernmental Panel on Climate Change, available at www.ipcc.ch (April 2007).
4. S. Pacala & R. Socolow, "Stabilization Wedges: Solving the Climate Problem for the Next 50 Years with Current Technologies," *Science*, **305**, 968-972 (August 13, 2004).
5. "International Energy Outlook 2010," Energy Information Administration, available at www.eia.doe.gov (March 2011).
6. "The Summer 2001 Conservation Report," available at www.energy.ca.gov (2007).

Chapter 2

1. "Annual Energy Review 2009," DOE/EIA-0384(2009), Energy Information Administration, available at www.eia.doe.gov (March 2011).

2. T.W. Patzek, "The Real Biofuel Cycles," University of California-Berkeley, available at petroleum.berkeley.edu/papers/Biofuels/MyBiofuelPapersTop.htm (March 26, 2006).
3. A.O. Alabi, M. Tampier & E. Bibeau, "Microalgae Technologies & Processes for Biofuels/Bioenergy Production in British Columbia," British Columbia Innovation Council, available at http://www.bcic.ca (January 2009).
4. H. Shapouri, J.A. Duffield & M. Wang, "The Energy Balance of Corn Ethanol: An Update," U.S. Department of Agriculture Report # 813 (July 2002).
5. T.W. Patzek, "Thermodynamics of the Corn-Ethanol Biofuel Cycle," University of California-Berkeley, available at petroleum.berkeley.edu/papers/Biofuels/MyBiofuelPapersTop.htm (July 22, 2006).
6. "Biofuels for Transport," International Energy Agency, available at www.iea.org/textbase/nppdf/free/2004/biofuels2004.pdf (March 2007).
7. D. Pimentel & T.W. Patzek, "Ethanol Production Using Corn, Switchgrass, and Wood; Biodiesel Production Using Soybean and Sunflower," *Natural Resources Research*, **14**(1), 65-76 (March 2005).
8. E. Smeets, M. Junginger & A. Faaij, "Sustainability of Brazilian Bio-Ethanol," Report NWS-E-2006-110 (August 2006).
9. T.W. Patzek, "The Earth, Energy, and Agriculture," Climate Change and the Future of the American West, available at petroleum.berkeley.edu/papers/Biofuels/MyBiofuelPapersTop.htm (June 7-9, 2006).
10. A.E. Farrell et al., "Ethanol Can Contribute to Energy and Environmental Goals," *Science*, **311**, 506-508 (January 27, 2006).
11. E. Rosenthal, "Once a Dream Fuel, Palm Oil May Be an Eco-Nightmare," *New York Times* (January 31, 2007).

References

Chapter 3

1. J.J. Romm, *The Hype about Hydrogen,* Island Press: Washington, D.C. (2004).
2. J.J. Romm, "The Car and Fuel of the Future," *Energy Policy,* **34**(17), 2609-2614 (August 2005).
3. U. Bossel, "Does a Hydrogen Economy Make Sense?" *Proceedings of the IEEE,* **94**(10), 1826-1837 (October 2006).
4. W.A. Summers et al., "Plant Definition and Economic Analysis for Centralized Nuclear Hydrogen Production," *Transactions of the American Nuclear Society,* **92**, San Diego, California (June 5-9, 2005).
5. M. Ehsani et al., *Modern Electric, Hybrid Electric, and Fuel Cell Vehicles: Fundamentals, Theory, and Design,* CRC Press LLC: New York (2005).

Chapter 4

1. J.J. Romm & A.A. Frank, "Hybrid Vehicles," *Scientific American,* **294**(4), 56-63 (April 2006).
2. M. Ehsani et al., *Modern Electric, Hybrid Electric, and Fuel Cell Vehicles: Fundamentals, Theory, and Design,* CRC Press LLC: New York (2005).
3. R. Cogan, "Quick Charge," *Green Car Journal* (on-line), available at www.greencar.com (2007).
4. "Toshiba's New Rechargeable Lithium-Ion Battery Recharges in Only One Minute," available at www.toshiba.co.jp (March 29, 2005).
5. "Altair Nanotechnologies Begins Manufacturing High-Power Lithium Ion Battery Cells," available at www10.mcadcafe.com (2007).
6. U. Bossel, "Does a Hydrogen Economy Make Sense?" *Proceedings of the IEEE,* **94**(10), 1826-1837 (October 2006).

7. E. Alsema & A. Patyk, "Investigation on Storage Technologies for Intermittent Renewable Energies: Evaluation and Recommended R&D Strategy," available at www.chem.uu.nl/nws/www/publica/Publicaties2003 (November 2003).
8. "Annual Energy Review 2009," DOE/EIA-0384(2009), Energy Information Administration, available at www.eia.doe.gov (March 2011).
9. J.J. Romm, "The Car and Fuel of the Future," *Energy Policy,* **34**(17), 2609-2614 (August 2005).
10. W.D. Jones, "Take this Car and Plug It," *IEEE Spectrum,* **42**(7), 10-13 (July 2005).
11. Nissan Official Site, available at www.nissanusa.com (March 2011).
12. Chevrolet Official Site, available at www.chevrolet.com/volt (March 2011).

Chapter 5

1. F. Trieb et al., "Solar Electricity Generation—A Comparative View of Technologies, Costs and Environmental Impact," *Solar Energy,* **59**(1-3), 89-99 (1997).
2. P. Breeze, *Power Generation Technologies,* Elsevier: Oxford, U.K. (2005).
3. National Renewable Energy Laboratory, available at www.nrel.gov/data/pix/ (May 2007).
4. T.B. Johansson et al., *Renewable Energy: Source for Fuels and Electricity,* Island Press: Washington, D.C. (1993).
5. Energy Information Administration, available at www.eia.doe.gov (July 2006).
6. P. Meier, "Life-Cycle Assessment of Electricity Generation Systems and Applications for Climate Change Policy Analysis," Dissertation, University of Wisconsin-Madison (August 2002).

7. F. Kreith, P. Norton & D. Brown, "A Comparison of CO_2 Emissions from Fossil and Solar Power Plants in the Unites States," *Energy*, **15**(12), 1181-1198 (1990).
8. "Levelized Cost of New Generation Resources in the Annual Energy Outlook 2011," available at www.eia.doe.gov (April 2011).
9. "Assessment of Parabolic Trough and Power Tower Solar Technology Cost and Performance Forecasts," NREL/SR-550-34440, Sargent & Lundy LLC Consulting Group (October 2003).
10. "Updated Capital Cost Estimates for Electricity Generation Plants," U.S. Energy Information Administration, available at www.eia.doe.gov (November 2010).
11. G. Barbose, N. Darghouth & R. Wiser, "Tracking the Sun III: The Installed Cost of Photovoltaics in the U.S. from 1998-2009," Lawrence Berkeley National Laboratory, LBNL-4121E (December 2010).

Chapter 6

1. D. Jennejohn, "US Geothermal Power Production and Development Update," Geothermal Energy Association, available at geo-energy.org (April 2010).
2. C.F. Kutscher, "The Status and Future of Geothermal Electric Power," NREL/CP-550-28204 (August 2000).
3. Energy Efficiency and Renewable Energy, U.S. Department of Energy, available at www.eere.energy.gov (April 2007).
4. "The Future of Geothermal Energy: Impact of Enhanced Geothermal Systems (EGS) on the United States in the 21st Century," INL/EXT-06-11746, Massachusetts Institute of Technology (2006).

5. "Enhanced Geothermal Systems," available at www1.eere.energy.gov/geothermal/egs_technology.html (January 2007).
6. P. Breeze, *Power Generation Technologies*, Elsevier: Oxford, U.K. (2005).
7. P. Berinstein, *Alternative Energy: Facts, Statistics, and Issues*, Oryx Press: Westport, Connecticut (2001).
8. T.B. Johansson et al., *Renewable Energy: Source for Fuels and Electricity*, Island Press: Washington, D.C. (1993).
9. J. McKenna et al., "Geothermal Electric Power Supply Possible from Gulf Coast, Midcontinent Oil Field Waters," *Oil and Gas Journal*, **103**(33), 34-40 (September 5, 2005).

Chapter 7

1. P. Breeze, *Power Generation Technologies*, Elsevier: Oxford, U.K. (2005).
2. "Projected Costs of Generating Electricity: 2005 Update," OECD Nuclear Energy Agency (2005).
3. "Ocean Energy Development: Obstacles to Commercialization," *Oceans 2003 Proceedings*, **4**, 2278-2283 (September 22-26, 2003).
4. T.B. Johansson et al., *Renewable Energy: Source for Fuels and Electricity*, Island Press: Washington, D.C. (1993).
5. A. von Jouanne, "Harvesting the Waves," *Mechanical Engineering*, **128**(12), 24-27 (December 2006).
6. Energy Efficiency and Renewable Energy, U.S. Department of Energy, available at www.eere.energy.gov (March 2007).
7. Pelamis Wave Power Ltd., available at www.pelamiswave.com (March 2011).
8. Oregon State University, available at eecs.oregonstate.edu/msrf (March 2007).
9. O. Siddiqui & R. Bedard, "Feasibility Assessment of Offshore Wave and Tidal Current Power Production:

A Collaborative Public/Private Partnership," 2005 IEEE Power Engineering Society General Meeting, **2**, 2004-2010 (June 12-16, 2005).
10. OCS Alternative Energy, available at ocsenergy.anl.gov (March 2007).
11. Marine Current Turbines TM Ltd., available at www.marineturbines.com (March 2007).

Chapter 8

1. "Annual Energy Review 2009," DOE/EIA-0384(2009), Energy Information Administration, available at www.eia.doe.gov (March 2011).
2. P. Berinstein, *Alternative Energy*, Oryx Press: Westport, Connecticut (2001).
3. P. Breeze, *Power Generation Technologies*, Elsevier: Oxford, U.K. (2005).
4. T.B. Johansson et al., *Renewable Energy: Source for Fuels and Electricity*, Island Press: Washington, D.C. (1993).
5. P.L. Spath & M.K. Mann, "Biomass Power and Conventional Fossil Systems with and without CO_2 Sequestration – Comparing the Energy Balance, Greenhouse Gas Emissions and Economics," NREL/TP-510-32575 (January 2004).
6. "Updated Capital Cost Estimates for Electricity Generation Plants," U.S. Energy Information Administration, available at www.eia.doe.gov (November 2010).
7. "Levelized Cost of New Generation Resources in the Annual Energy Outlook 2011," available at www.eia.doe.gov (April 2011).
8. M. Varela et al., "Large-Scale Economic Integration of Electricity from Short-Rotation Woody Crops," *Solar Energy*, **70**(2), 95-107 (2001).

Chapter 9

1. T.B. Johansson et al., *Renewable Energy: Source for Fuels and Electricity*, Island Press: Washington, D.C. (1993).
2. Energy Information Administration, available at www.eia.doe.gov (March 2011).
3. "Wind Power Today," U.S. Department of Energy, DOE/GO-102006-2319 (May 2006).
4. P. Breeze, *Power Generation Technologies*, Elsevier: Oxford, U.K. (2005).
5. J.F. DeCarolis & D.W. Keith, "The Economics of Large-Scale Wind Power in a Carbon Constrained World," *Energy Policy*, **34**, 395-410 (2006).
6. R.Y. Redlinger et al., *Wind Energy in the 21st Century*, Palgrave: New York (2002).
7. International Energy Agency, available at www.iea.org (April 2007).
8. S. White & G.L. Kulcinski, "Birth to Death Analysis of the Energy Payback Ratio and CO_2 Gas Emission Rates from Coal, Fission, Wind, and DT-Fusion Electrical Power Plants," *Fusion Engineering and Design,* **48**, 473-481 (2000).
9. "Updated Capital Cost Estimates for Electricity Generation Plants," U.S. Energy Information Administration, available at www.eia.doe.gov (November 2010).
10. "Levelized Cost of New Generation Resources in the Annual Energy Outlook 2011," available at www.eia.doe.gov (April 2011).
11. "Projected Costs of Generating Electricity: 2005 Update," OECD Nuclear Energy Agency (2005).
12. W. Kempton et al., "The Offshore Wind Power Debate: Views from Cape Cod," *Coastal Management*, **33**, 119-149 (2005).

References

Chapter 10

1. Energy Information Administration, available at www.eia.doe.gov (March 2011).
2. C.E. Brown, *World Energy Resources*, Springer: Berlin, Germany (2002).
3. R.G. Cochran & N. Tsoulfanidis, *The Nuclear Fuel Cycle: Analysis and Management*, American Nuclear Society: Lagrange Park, Illinois (1990).
4. G. Petrangeli, *Nuclear Safety*, Elsevier: Oxford, England (2006).
5. "Nuclear News Special Report: Fukushima Daiichi after the Earthquake and Tsunami," available at www.new.ans.org/pubs (April, 2011).
6. T.L. Schulz, "Westinghouse AP1000 Advanced Passive Plant," *Nuclear Engineering and Design*, **236**(14-16), 1547-1557 (2006).
7. American Nuclear Society, available at www.ans.org (February 2007).
8. Nuclear Regulatory Commission, available at www.nrc.gov/reading-rm/doc-collections/fact-sheets/3mile-isle.html (June 2007).
9. S. Hirschberg, G. Spiekerman & R. Dones, "Severe Accidents in the Energy Sector," Paul Scherrer Institut, ISSN-1019-0643 (November 1998).
10. D.E. Shropshire et al., "Advanced Fuel Cycle Cost Basis," INL/EXT-07-12107 (March 2007).
11. "Final Environmental and Impact Statement for a Geological Repository for the Disposal of Spent Nuclear Fuel and High-Level Radioactive Waste at Yucca Mountain," DOE/EIS-0250, U.S. Department of Energy, Office of Civilian Radioactive Waste Management (February 2002).
12. S. White & G.L. Kulcinski, "Birth to Death Analysis of the Energy Payback Ratio and CO_2 Gas Emission Rates from Coal, Fission, Wind, and DT-Fusion

Electrical Power Plants," *Fusion Engineering and Design,* **48**, 473-481 (2000).
13. "Updated Capital Cost Estimates for Electricity Generation Plants," U.S. Energy Information Administration, available at www.eia.doe.gov (November 2010).
14. "Levelized Cost of New Generation Resources in the Annual Energy Outlook 2011," available at www.eia.doe.gov (April 2011).

Chapter 11

1. T.K. Fowler, *The Fusion Quest,* Johns Hopkins University Press: Baltimore, Maryland (1997).
2. "ITER Technical Basis," available at www.iter.org (February 2007).
3. G.H. Miller, E.I. Moses & C.R. Wuest, "The National Ignition Facility: Enabling Fusion Ignition for the 21st Century," *Nuclear Fusion,* **44**(12), S228-S238 (2004).
4. R.L. Hirsch, "Fusion Power: The Burning Issue," *Public Utilities Fortnightly* (February 1, 2003).
5. L.M. Lidsky, "The Trouble with Fusion," *Technology Review,* **32**, (September 1983).
6. F. Najmabadi et al., "Overview of the ARIES-RS Reversed-Shear Tokamak Power Plant Study," *Fusion Engineering and Design,* **38**, 3-25 (1997).
7. F. Najmabadi et al., "Spherical Torus Concept as Power Plants—the ARIES-ST Study," *Fusion Engineering and Design,* **65**, 143-164 (2003).
8. F. Najmabadi et al., "The ARIES-AT Advanced Tokamak Advanced Technology Fusion Power Plant," *Fusion Engineering and Design,* **80**, 3-23 (2006).

References

Chapter 12

1. P. Breeze, *Power Generation Technologies*, Elsevier: Oxford, U.K. (2005).
2. G.R. Couch, "Competitive Position of Coal for Power Generation," ISBN 92-9029-399-3, IEA Clean Coal Centre (May 2004).
3. C. Henderson, "Towards Zero Emission Coal-Fired Power Plants," ISBN 92-9029-417-5, IEA Clean Coal Centre (September 2005).
4. C. Henderson, "Clean Coal Technologies," ISBN 92-9029-389-6, IEA Clean Coal Centre (October 2003).
5. D.J. Borns et al., "Carbon Sequestration and Clean Coal Technologies: Characterizing Systems and Evaluating Costs," SAND 2005-5917P (September 2005).
6. "Clean Coal Technology Roadmap," available at www.coal.org (October 2006).
7. Z. Wu, "Air Pollution Control Costs for Coal-Fired Power Stations," ISBN 92-9029-366-7, IEA Clean Coal Center (October 2001).
8. S. White & G.L. Kulcinski, "Birth to Death Analysis of the Energy Payback Ratio and CO_2 Gas Emission Rates from Coal, Fission, Wind, and DT-Fusion Electrical Power Plants," *Fusion Engineering and Design*, **48**, 473-481 (2000).
9. "Updated Capital Cost Estimates for Electricity Generation Plants," U.S. Energy Information Administration, available at www.eia.doe.gov (November 2010).
10. "Levelized Cost of New Generation Resources in the Annual Energy Outlook 2011," available at www.eia.doe.gov (April 2011).
11. CBS News/New York Times Poll, available at www.pollingreport.com, (April 20-24, 2007).

Chapter 13

1. "Annual Energy Review 2009," DOE/EIA-0384(2009), Energy Information Administration, available at www.eia.doe.gov (March 2011).
2. "Annual Energy Outlook 2010," DOE/EIA/0383 (2010), Energy Information Administration, available at www.eia.doe.gov (March 2011).
3. P. Breeze, *Power Generation Technologies*, Elsevier: Oxford, U.K. (2005).
4. Energy Information Administration, available at www.eia.doe.gov (March 2011).
5. P.L. Spath & M.K. Mann, "Life Cycle Assessment of a Natural Gas Combined-Cycle Power Generation System," NREL/TP-570-27715 (September 2000).
6. "Updated Capital Cost Estimates for Electricity Generation Plants," U.S. Energy Information Administration, available at www.eia.doe.gov (November 2010).
7. "Levelized Cost of New Generation Resources in the Annual Energy Outlook 2011," available at www.eia.doe.gov (April 2011).
8. C. Henderson, "Towards Zero Emission Coal-Fired Power Plants," ISBN 92-9029-417-5, IEA Clean Coal Centre (September 2005).
9. "Assumptions to the Annual Energy Outlook 2006," DOE/EIA-0554, Energy Information Administration (March 2006).
10. "Projected Costs of Generating Electricity: 2005 Update," OECD Nuclear Energy Agency (2005).

About the Author

Ben Cipiti received his bachelor's degree in mechanical engineering from Ohio University and PhD in nuclear engineering from the University of Wisconsin-Madison. He currently works at Sandia National Laboratories in Albuquerque, New Mexico, with research interests in energy economics, fusion energy, nuclear waste reduction, and nuclear material safeguards.

St. Louis Community College
at Meramec
LIBRARY

Made in the USA
Lexington, KY
10 November 2012